Insights into Teaching Mathematics

Anthony Orton and Leonard Frobisher

CASSELL

Cassell
Wellington House
125 Strand
London WC2R 0BB

215 Park Avenue South
New York, NY 10003

First published 1996

British Library Cataloguing-in-Publication Data
A catalogue record for this book is available from the British Library.

ISBN 0–304–33218–6 (hardback)
    0–304–33220–8 (paperback)

Typeset by Action Typesetting
Printed and bound in Great Britain by Redwood Books Limited, Trowbridge, Wiltshire.

INTRODUCTION TO EDUCATION

Series editor: Jonathan Solity

INSIGHTS INTO TEACHING MATHEMATICS

BOOKS IN THIS SERIES

Contents

Foreword

The 1980s and 1990s have witnessed unprecedented changes to the education system. These have had a dramatic impact, particularly in relation to:

- schools' relationships with parents and the community;
- the funding and management of schools;
- the curriculum;
- the assessment of children's learning.

It can be an extremely daunting task for student teachers to unravel the details and implications of these initiatives. This Introduction to Education series therefore offers a comprehensive analysis and evaluation of educational theory and practice in the light of recent developments.

The series examines topics and issues of concern to those entering the teaching profession. Major themes representing a spectrum of educational opinion are presented in a clear, balanced and analytic manner.

The authors in the series are authorities in their field. They emphasize the need to have a well-informed and critical teaching profession and present a positive and optimistic view of the teacher's role. They endorse the view that teachers have a significant influence over the extent to which any legislation or ideology is translated into effective classroom practice.

Each author addresses similar issues, which can be summarized as:

- presenting and debating theoretical perspectives within appropriate social, political, and educational contexts;
- identifying key arguments;
- identifying individuals who have made significant contributions to the field under review;
- discussing and evaluating key legislation;
- critically evaluating research and highlighting implications for classroom practice;

- providing an overview of the current state of debate within each field;
- describing the features of good practice.

The books are written primarily for student teachers. However, they will be of interest and value to all those involved in education.

Jonathan Solity
Series Editor

Preface

This book aims to provide a brief introduction to many of the essential issues relating to the teaching of mathematics. It is aimed particularly at satisfying the needs of those new teachers and trainee teachers who do not consider themselves to be mathematics specialists. However, it should also provide valuable background information and guidance for all who teach mathematics across the primary and middle years of schooling. The main concern is with children in the age range 5–14 years, and therefore issues and content of more relevance to older pupils and students have not been included. We believe that the book is comprehensive, in the sense that it not only deals with all the main contemporary issues relating to mathematics teaching, but also provides practical advice and assistance for teaching all the main strands into which the mathematics curriculum is often divided. We have tried to approach the task of giving direct teaching advice from the standpoint of relevant research findings concerning how pupils learn and the difficulties they often seem to encounter. If we have omitted to include matters which you hoped or expected to find here, this has been done deliberately in order to produce a relatively short book. On the other hand, it is likely that not all parts of the chapters on content will be relevant to the particular children you teach. We would like to acknowledge the valuable help given to us by Jean Orton, in reading, checking and commenting on our work, but we take full responsibility for the entire contents of the book.

Tony Orton
Len Frobisher
Leeds, 1995

CHAPTER 1

The mathematics curriculum

MATHEMATICS AND THE SCHOOL CURRICULUM

Mathematics is widely regarded as one of the most important subjects in the school curriculum. Indeed, it is likely that more lessons of mathematics are taught in schools and colleges throughout the world than any other subject. In the United Kingdom it is necessary to understand that three separate mathematics curricula exist, namely those for England and Wales, Scotland and Northern Ireland. In England and Wales, the National Curriculum has always defined mathematics, along with English and science, as a 'core' subject. Each of these three subjects is expected to occupy most children for more teaching time than any other subject. When concern is expressed about the attainment of pupils in England and Wales and comparisons, whether legitimate or not, are made with pupils in other countries, mathematics is usually singled out as being a particularly worrying problem. In fact, it must be made clear that there are very many countries in the world where great concern is frequently expressed about attainment in mathematics. It seems that the whole world regards it as important that children should be able to demonstrate a high level of competence in the subject.

The importance of mathematics is further emphasized when the future employment of a child is being considered, with parents almost unanimously wanting their children to succeed in the subject, largely in the hope that job prospects will be improved. What is more, mathematics is used as a 'filter' or 'hurdle' possibly more often than any other subject, in that an examination pass at an appropriate level is demanded before entry to a particular profession or occupation can even be considered – whether any mathematics is required in the performance of the job or not. In England and Wales, even those teachers who will never teach any mathematics currently need to be able to produce evidence of satisfactory attainment in mathematics.

One reason why mathematics is used as a filter is that it is often claimed to be equated with clear thinking and the ability to solve problems. The assumption is frequently made that, in order to produce a new generation of adults who are better at solving the problems of science, life, the world, and whatever, we need to strive for higher attainment in mathematics from greater numbers of pupils. In fact, there is no convincing evidence that learning mathematics improves thinking and problem-solving skills *per se*, and we should not assume that mathematics has more to offer than other subjects in the curriculum in relation to learning how to solve problems. It is highly questionable whether mathematics is any more important than any other subject, never mind whether it is equally important to the future of every child, by whatever criteria are chosen to measure importance. Nevertheless, those of us who teach the subject do so against this background of high regard by society for the importance of mathematics.

This seemingly internationally ubiquitous elevated position of mathematics, and of those who succeed in the subject, has unfortunate consequences both for individual learners and for mathematics. As far as individual children are concerned, great pressures to succeed are often placed on them, and this inevitably makes it even more difficult to be successful. All of us are likely to discover that our performance is worse and our success rate is lower when we are under pressure to succeed and, what is more, we are also likely to develop a negative attitude towards what we are expected to do. There is no denying that mathematics is essentially an abstract subject and, despite valiant attempts by generations of teachers to improve the quality of pupils' learning, abstractions remain difficult to grasp. No matter how far any one of us has progressed with mathematics, at one time or another all but a very few of us have experienced difficulties in understanding particular abstract ideas. Perhaps we need to convey this message to pupils, in order to alleviate some of the pressures which many pupils and students undoubtedly feel. Pressure to succeed and to pass examinations obviously makes it more difficult for learners to be at ease with the subject matter, to develop a state of mind which is receptive to the idea that mathematics can be enjoyable and need not generate anxiety and panic. Pressure also makes it more difficult for teachers of mathematics at any level to aim to teach for the enjoyment of learning mathematics rather than future examination success. A major assumption of this book is that mathematics can and should be enjoyed by both teacher and pupil, and that quality of learning can thus be improved whilst what society regards as important and essential need not suffer.

Attitudes adopted by society as a whole go far beyond those expressed previously. The high esteem in which mathematics is held has led to the view that those who are good at mathematics are particularly intelligent. At the same time, those who profess not to be any good at mathematics do so all too frequently and almost with pride. Perhaps this is a claim to normality, with those who are successful in the subject being regarded as the abnormal minority, particularly if they are female. Whatever the reason, it is not helpful for the future of our children and it reflects badly upon school mathematics that so many people seem to take pleasure in claiming they were hopeless at mathematics when at school, especially when it seems that similar claims about other subjects are made much more rarely. There are even many adults who achieved examination success in mathematics at the age of sixteen years, and even at eighteen years, who still claim not to be any good at the subject.

The greatest harm perpetrated by this widely expressed view is probably done to girls and, not surprisingly, often by their mothers. If society needs better mathematically educated adults, and if genuine equality of opportunity is to be provided, more girls need to be persuaded to consider studying mathematics for a longer period of time. There is no convincing evidence that girls are less capable than boys, so we need to do all we can to eliminate any harmful influence. Society needs to display attitudes to mathematics which do not predispose any group, be they adults or children, to adopt unjustifiable views. Above all, we need a curriculum, including methods of teaching and forms of examining, which allows children both to enjoy and succeed in the subject. This book is aimed particularly at teachers who may not regard themselves as mathematics specialists but who share these objectives.

THE AIMS OF TEACHING MATHEMATICS

Reasons why mathematics is universally agreed to be a vital area of the school curriculum have already been expressed, namely its assumed importance in life and its value in teaching people to think and solve problems. Neither of these justifications for teaching mathematics should be accepted uncritically, and they need to be considered separately. We also ought to consider whether there are other aims of teaching mathematics, which might even be more important or significant than the two which have arisen so far.

It is true that we all have the opportunity to use some mathematics in our daily lives, but most of us either do not use very much, or fail to recognize when we have done so. Many people rarely need more than the ability to add and subtract, and possibly multiply and divide a little, with relatively small numbers, and mostly in the context of handling money. As long ago as 1960, Dienes wrote, 'if the requirements of everyday life determined the contents of our mathematics syllabuses there would surely be little mathematics in them' (1960, p. 9). It is little consolation that the same could probably be said about any subject in the curriculum. In employment, some people do use some mathematics, but a great deal of what formerly required mental or written arithmetic on the job is now automated and depends more on pressing the correct keys than applying taught mathematical procedures or skills. (We shall take a procedure to be a step-by-step sequence of actions, physical or mental, a routine to be a procedure which is executed automatically, and a skill to be the ability to select an appropriate routine and apply it successfully.) As regards industry and business, the evidence from Fitzgerald (1981) is that the methods taught in school are often not used on the shop floor. This seems to suggest that the particular arithmetical procedures we teach in school may not be particularly important in later life. However, there is a strong argument that those who press keys on machines should have enough mathematical knowledge to know whether outcomes or answers are likely to be correct. Moreover, it is important for many people to know which mathematical procedures to activate through the available technology. It is also true that mathematical methods are used in a wide variety of other subject studies, but here we find that teachers of other subjects often themselves prefer to teach the mathematics which is needed, to some extent because the standard procedures which mathematicians teach are not the ones needed. The small amount of mathematics which is fundamental to all still needs to be taught, and this is an important responsibility of primary school teachers in particular. Beyond that, a major responsibility must be to aim to produce people who can use what is perhaps a limited amount of mathematics appropriately in all their thinking. This is a different aim from that of producing pupils who have been taught a large number of procedures which they may never need and which experience suggests they quickly forget.

There is no convincing evidence that studying mathematics is the best way to 'train the mind', or that a heavy diet of school mathematics produces better logical thinkers, or that any particular mathematical content is as valuable as any other. Mathematicians might well be able to call upon appropriate mathematical methods to solve some of life's problems, but this does not mean that, outside mathematics, they will solve non-mathematical problems better than other people. The history of philosophy and psychology reveals a view, quite widely held for a time, that doing mathematics stretches or exercises the mind, thus enabling us to think more clearly and successfully in any field of mental activity, rather in the same way that physical exercise might make us better able to take part in athletic pursuits of any kind. This view is not now accepted, and we should not attempt to justify teaching mathematics on such grounds. What is more, we should at the very least be suspicious when mathematics tests are used as a means of selecting individuals for occupations or pursuits in which mathematics is not required. However, nor must we swing to the other extreme of regarding mathematics as irrelevant. Learning mathematics is the only way of learning mathematical ways of thinking, and learning how mathematics might or might not be able to contribute to solving a particular problem. A curriculum based on this premise might be very different from one based on the assumption that any kind of mathematics will do because whatever it is it will teach us to think more clearly and successfully.

There are other reasons for teaching mathematics, for example the fact that our pupils have the right to know how important mathematics is in our modern world. Mathematics underpins science and technology, enabling banks and insurance companies to function efficiently just as much as it enables us to place artificial satellites precisely so as to improve communications round the world. There are so many applications of mathematics in our world that it is impossible even to begin to list them. What is clear, however, is that the present mathematics curriculum does not seem particularly effective in conveying this aspect of the importance of mathematics to our pupils. It could be argued that it is more important to teach this point effectively than to teach a collection of meaningless procedures such as long division.

Mathematics is also part of our cultural heritage, just as much as the plays of Shakespeare and the music of Mozart. In the struggles to master and control our environment, people down the ages have called upon, or even invented, mathematical procedures and other methods to enable better land measurement, navigation, transport, communication and, unfortunately, more efficient warfare. Brief snippets concerning the history of mathematics not only are relevant within mathematics lessons, but may be important and are often very much appreciated by the children. It should also be said, of course, that information about contemporary developments is equally important, so that children come to realize that mathematics is alive and constantly growing. And, if language is what enables us to communicate, then mathematics also provides us with extensions of language. Information is often best conveyed by means of tables and graphs, algebraic symbols provide us with a concise means of summarizing relationships such as in a formula, and diagrams often allow easier ways of conveying ideas than any other means.

Finally, like other subjects such as art and music, mathematics can be appreciated and enjoyed for its own sake. It is likely there is an aesthetic element to all subjects, but some subjects seem to spring more readily to mind for most people when aesthetics are being discussed, and mathematics is seldom considered at all under this heading. This is a pity, because there is most certainly an aesthetic element to mathematics, and if few people realize this, it may be because of what and how we teach. We owe it to our pupils to try to help them to enjoy and appreciate mathematics just as much as any other subject. Some people would say this should be our first aim, not an afterthought.

ATTAINMENT IN MATHEMATICS

It is not new for concern to be expressed about falling standards in education; such anxiety has a long history. Indeed, it was against a background of unsubstantiated accusations of unsatisfactory mathematics teaching in schools and suspicions of falling standards in Britain that the Cockcroft Committee was set up. The Cockcroft Report (1982) remains a very important document, and one which incidentally did not confirm that standards had fallen. If at any time it can be proved that standards are falling, then there is certainly a problem to be addressed. In fact, there are two vitally important questions which need to be answered on this issue. We need to ask what we mean by standards, and we also need to ask what is the evidence concerning whether our pupils know less and perform worse than they did in some previous golden age.

It is too easily assumed that there is universal agreement about what is meant by standards in education when there might not be. One definition held by some people demands that children in every generation should be able to perform the same tasks with the same degree of proficiency. This ignores the issue as to whether we should be asking the same questions now as we asked of former generations. It is certainly no longer relevant to ask questions involving pounds, shillings and pence (what were they, some readers may ask?), and most people would say the same about common logarithms. It is also not normal to ask questions involving the derivation of square roots by pencil and paper or even to require the proof of a geometrical theorem. Complicated sums involving fractions are not now generally part of mathematics practice in Britain, so it should not be a surprise to find that children are not as proficient with them as they were thirty or fifty years ago. Does the fact that children are less skilled in certain areas of mathematics mean that standards have fallen? We now devote less time in mathematics teaching to children acquiring proficiency in answering certain kinds of question, but we have introduced new topics into the curriculum, like probability and statistics, transformation geometry and spreadsheets. How can we measure whether standards have fallen when some topics have been omitted from the curriculum and new ones introduced? Particular groups in society, such as university mathematics professors, might legitimately express great concern about curriculum changes which have affected their work (Barnard and Saunders, 1994), and it is undeniably true that the school mathematics curriculum of today contains less algebra for prospective university mathematics and science students than it used to, as well as other omissions, but does this necessarily represent a drop in standards in the overall mathematical education experienced by pupils? For example, how should we adjust our expectations of pupils given that part of their mathematics course is now devoted to investigating, solving problems which claim to model real life, and using and applying some of the skills and techniques they have developed? If we accept the view expressed earlier that education is a preparation for life, for the ways in which our pupils will use mathematics in their daily lives and at work, and if the mathematics which we supposedly need has changed, it must be the case that school mathematics should also have changed to reflect what is now required. If we prefer certain other views about the purposes of education, such as producing well-educated individuals, then there may be even less reason to demand that the present curriculum should necessarily be exactly the same as in the past, when numerate clerks were required in comparatively large numbers. If the mathematics curriculum is not the same, how can we reasonably demand that pupils should be able to answer all the same mathematics questions as their parents? It may, however, still be the case that today's pupils should be able to answer some of the questions which their parents were expected to answer. The scope for disagreement about what those questions should be is enormous.

Whatever detailed evidence of changes in pupil attainment is available, we must consider seriously the issues just raised. If we wish to compare children of today with children of ten, twenty or fifty years ago, how do we take account of changes in the mathematics curriculum and differences in teaching emphasis and style? It may be possible for us to state only that children of today are better at some things but worse at others. If there is no evidence of improvement in any direction to compensate for deterioration in others, then a claim that standards have fallen might be justifiable. Evidence from the Assessment of Performance Unit (SEAC, 1988), reporting on changes over the five-year period 1982 to 1987, shows that children improved in some areas and slipped back in others. They were better at probability, geometry, measures, the algebra of generalized arithmetic and number concepts, and they were worse at concepts of fractions and decimals, computation skills and number applications, ratio, trigonometry and some aspects of algebra. It is noteworthy that the areas where there was some deterioration are largely related to proficiency with number, the part of the curriculum in which calculators have become much more widely used, sometimes with the express purpose of removing the drudgery of long and tedious calculations. After all, there are certain things which machines do better than humans, thus freeing active and creative minds for more original pursuits. These areas of relatively minor deterioration are also those which traditionalists will inevitably focus on in order to create anxiety and concern amongst the general public. Unfortunately, the APU no longer operates, so there will be no further evidence from that quarter, thus perhaps making accusations of falling standards more difficult to refute.

It is interesting that teachers, curriculum developers and examiners in Britain have recently responded very successfully to a demand that standards in our external examinations at age 16+ should be improved, and the outcome has been a higher pass rate. Paradoxically, this has led to the accusation that standards must have dropped, because more pupils are successful! The concept of standards in education is an elusive one indeed. International comparisons (Cresswell and Gubb, 1987; Robitaille and Garden, 1989), have found it equally difficult to come to any conclusions of value, owing to difficulties of carrying out testing in a variety of countries with different educational systems, ages of schooling, and curriculum, as well as diversity of cultural tradition and attitudes of society. On the surface, it appears that British children are fairly average, comparatively better than children in many other countries at some things and worse at others. Children in certain other countries, particularly in the Far East, often seem much better than our children, particularly in arithmetic, but cultural traditions and parental and societal expectations appear to offer some explanation for this phenomenon. The maintenance of satisfactory standards, indeed the improvement of standards, is an issue which has to be taken seriously, but common sense suggests that much has improved throughout this century, and continues to improve. There is no room for complacency, but evidence needs to be collected and analysed carefully, and simplistic conclusions based on suspect evidence do not help us to monitor progress.

THE NATIONAL CURRICULUM

There is more than one national curriculum (NC) in the United Kingdom, because of the separate arrangements for England and Wales, Scotland and Northern Ireland. The situation is further complicated in that independent schools are not bound by the requirements of any NC. The main difficulty associated with commenting on these curricula is that there have already been a number of quite radical changes, even since their first introduction in 1989. Indeed, British teachers have for some years looked forward to a period of greater stability, in order that changes could be accommodated effectively and developments consolidated. The authors of many school textbooks published in recent years have fallen foul of this phenomenon of rapid change in the documentation written to guide teachers. Here, we shall attempt to avoid that particular problem, but it remains to be seen whether we have succeeded.

Mathematics being accorded the status of a 'core' subject in the NC, and consequently occupying a considerable amount of a pupil's time, is not a new concept in British education. In fact, it is likely that mathematics will always be regarded as a core subject all around the world. For the purposes of defining the subject matter to be taught, mathematics in the NC was, from the outset in 1989, subdivided into a number of domains, described as Attainment Targets. This terminology immediately makes crystal clear that the measurement of attainment is at the heart of the purposes of the NC, which brings us back to the issue of standards and their maintenance. What is more, the NC is enshrined in law, implying that any teacher refusing to adopt it stands the risk of being accused of committing a criminal offence.

In 1989, there were fourteen of these Attainment Targets, but they were reduced to five within three years of the introduction of the NC, presumably at least partly because fourteen was thought to be unnecessarily complicated. After all, the more subdivisions there are, the more difficulties there are in deciding what goes in each. In the end, many would say that mathematics should be regarded as a unity, so any subdivisions should be only those which are helpful or can be justified as necessary. Thus the five new areas defined in 1991 corresponded to the four 'traditional' subdivisions of mathematics into arithmetic (now usually called 'number'). algebra, geometry and probability with statistics, together with the fifth domain, a new one entitled 'using and applying mathematics'. The latest version of the curriculum has only four, with number and algebra being combined into one. This change, which might appear trivial in terms of curriculum partitioning, does offer the benefit of making more equal both the time required for teaching each of the four areas at all age levels, and the coverage of the curriculum areas in national tests. In number and algebra, the emphasis is naturally expected to be on number in the early years, and on algebra in the later years. There has been considerable debate over recent years about 'using and applying mathematics', and whether what is required of teachers should be integrated into the content-based areas. The argument for integration is that 'using and applying' cannot be taught as a separate subject area, for it is by definition concerned with using and applying the mathematics of all the other content areas. The argument for separation is that only by defining separately what it is that teachers must teach, and seeking to assess what children achieve in this area, will it be taught thoroughly and carefully, and will we be able to find out what has been achieved. This debate is set to continue into the future, whatever decisions are made about its place within the curriculum at any moment in time. In considering the content of the mathematics curriculum within this book, the 'traditional' division referred to above has been used.

ISSUES OF CURRICULUM CONSTRUCTION

Nothing was said in the previous section about teaching method. In fact, there are many issues of curriculum construction and definition, apart from content or subject matter, and some of these issues are considered here. In 1989, when the first NC was published, guidance and suggestions for teaching methods were not included. This might have been because of the great haste with which the curriculum was introduced, for it is clear that many of those people who have pressed for change in education in recent years have held strong views on methodology as well as content. The expressed claim, however, has often been that decisions about methodology should be left to the individual classroom teacher, taking into account the needs of their pupils and the purposes of particular lessons. Methods of teaching were discussed at length in the Cockcroft Report (1982, p. 71), where it was suggested that:

> Mathematics teaching at all levels should include opportunities for
> exposition by the teacher;
> discussion between teacher and pupils and between pupils themselves;
> appropriate practical work;
> consolidation and practice of fundamental skills and routines;
> problem solving, including the application of mathematics to everyday situations;
> investigational work.

This still stands as the best short statement of advice available. The suggestion has also been made frequently, both before and after the introduction of the NC, that mathematics teachers as a whole have traditionally been good at exposition, and at teaching aimed at consolidation and practice of skills and routines, but much less good at the other four elements of the Cockcroft recommendations. It is therefore unfortunate that the NC does not appear to lend much positive support to the wider range of Cockcroft suggestions. A document entitled 'Mathematics: Non-Statutory Guidance' was issued alongside the NC in 1989. It contains some useful amplification in such areas as progression and continuity, using and applying mathematics and mathematics across the curriculum, but it is non-statutory, and therefore can presumably be ignored by teachers.

The origins of the Cockcroft list of teaching methods are probably many and varied, but one is probably the long-standing debate about the nature of mathematics. What precisely is mathematics? Is it just subject matter to be learned or mastered, with the school mathematics curriculum being a small and elementary subset of the totality of all known mathematics, or are there alternative views? In recent years, a view has emerged that mathematics is 'process' as well as subject content (sometimes rather confusingly referred to as 'product'), that learning mathematics must involve learning or experiencing the 'processes' of mathematics. We suggest that processes, in this context, include the selection, assembly and application of knowledge and skills appropriate to the situation or task. Thus, basically, it is suggested that learning mathematics should involve learning something of what it is to be a mathematician. For this reason, investigating and solving non-routine problems in mathematics have become important, and in order to investigate or solve problems it is necessary to develop such process skills as conjecturing, collecting data systematically, analysing, generalizing and proving. Irrespective of other justifications for fostering discussion in mathematics lessons – and some such justifications will be considered in the next chapter – it should be clear that discussion can help in investigating and proving. There are often times in life when two or more minds might well be better than one, though whenever this is used to justify cooperation between pupils it is always necessary to try to ensure that individual children do not opt out and leave the thinking to others. Although it will sometimes be necessary to find out what an individual child is capable of, there is some justification for allowing children to work together at appropriate times in order to enhance learning. If one believes that children must 'do' mathematics, as well as practise appropriate methods, techniques and skills, then one can understand the relevance of the full Cockcroft list.

It is then also a natural step to accept the inclusion of 'using and applying' mathematics as an important component of the NC. The Cockcroft list includes reference to 'the application of mathematics to everyday situations', and therefore seems to imply that some problems to be solved will not concern everyday situations, thus suggesting that when we use and apply our mathematics, in the context of the NC, we will sometimes be using and applying it in connection with everyday situations and sometimes not.

Another important issue which should be raised here is the extent to which the use of calculators and computers can be justified in mathematics teaching, and how the curriculum should respond to their existence in the classroom. The introduction of cheap electronic calculators into schools in the 1980s opened up yet another debate, and one which still continues. Calculators can perform calculations more quickly and, unless incorrect keys are pressed, more accurately than people. What is more, there is no convincing evidence that calculators generate more errors than pencil-and-papers methods, and it is in fact more likely that they generate fewer. Calculators can help children to understand the way numbers work; they can be used to expose the structures of our number system, often more clearly than pencil and paper arithmetic can, simply because they reduce complexity and free mental processing space so that the brain can concentrate the better on ideas rather than on accuracy. Calculators can enable mathematics to be done in new, and sometimes simpler, ways, and can open up new areas of the subject which were previously inaccessible because of the complexity of the arithmetic. The same can be said of computers. In fact, any claims in favour of calculators probably apply even more strongly to computers. The initial worry about calculators, which for some people has never gone away, is that children will not learn essential number skills if they are allowed to use calculators too soon. However, children meet calculators and computers in their world outside school, not least quite likely in their homes, and calculators and computers are ubiquitous in modern society, to such an extent that to ignore them at school would lead to an accusation that we were not preparing children for the real world.

Calculators and computers cannot be ignored. The teaching of mathematics has to accommodate them, and much has been learned in recent years as to how best to use them in education, whilst still ensuring that children continue to develop basic skills in mental mathematics. The PrIME Project (see Shuard *et al.*, 1991) provides convincing evidence that the incorporation of calculators into mathematics lessons can lead to improvement in the mathematical education of pupils and in their attainment.

CHAPTER 2

Learning mathematics

TEACHING AND LEARNING

We teach in order that others may learn. Unfortunately, although they *may* learn, they also may not. We must, of course, acknowledge that a great deal of learning takes place without explicit teaching, through independent observation, reading, study and reflection, for example, and that school is only one of many environments in which children learn. It is, however, probably the case that most of the *mathematics* that children learn is encountered in school. Indeed, it might be that greater success in learning mathematics would be achieved if ideas from the subject featured rather more in conversation between adults and between parents and their children, just as aspects of many other subjects so often do. The fact that learning takes place outside school is only rarely a problem, but the converse is serious, namely that learning does not always take place inside school. Anyone who has attempted to teach, at any level, is aware that not all learners develop their capabilities in the way or to the extent that was hoped, and that some learners benefit more than others from any learning experience. This immediately raises questions as to whether adopting different teaching methods, such as a wider variety than those advocated in the Cockcroft Report (1982) and listed in the previous chapter, would lead to greater success, or whether there are other reasons for limitations in the learning which takes place.

Traditionally, mathematics teaching has relied heavily on exposition by the teacher together with the consolidation and practice by the learner of fundamental skills and routines, though the younger the child the less likely that exposition is the main method used, and the more likely that practical work with concrete objects will dominate. Superficially, exposition seems to be a very efficient method, with large numbers of pupils processed all at the same time and in the same way. Progress through the curriculum appears to be very much under the control of the teacher, and careful planning ensures the 'delivery' of the entire curriculum. What is more, exposition often seems to be economical in terms of effort. Much of what is transmitted consists of routines, such as how to add, subtract, multiply and divide two fractions, and it seems obvious that children must practise such routines in order to reinforce what has been transmitted. The main problem is the one expressed earlier, namely that every teacher of mathematics has encountered children for whom repeated exposition and practice of the same routine never leads to mastery, at least not within the school environment. Many procedures are not remembered correctly, if they are remembered at all, and sometimes procedures such as those for adding and multiplying fractions become confused. The contemporary suggestion that mathematics can be 'delivered' to children is as meaningless as it is common nowadays, unfortunately, and even practice cannot compensate for the shortcomings of delivery. It is essential that the other suggestions in the Cockcroft list are treated seriously.

The underlying issues, here, are those of how learning actually takes place, together with the nature of the kind of learning we wish to promote and what teaching methods best encourage learning. How mathematics is learned has been the subject of study and research by mathematicians, psychologists and learning theorists for more than a century, possibly much longer. Contrasting theories have been propounded, though at present there is considerable consensus that exposition and practice alone are not sufficient to promote learning most effectively. This is not to say that such 'traditional' methods may not sometimes be appropriate, or even necessary, but they do not have to be the only methods used.

The use of exposition and practice undoubtedly became established in our schools with the development of mass education, perhaps at least partly because these methods were seen as an obvious way of coping with large numbers of children, and not because they were known to be effective under all circumstances. To some extent, they were supported by behaviourist psychology, which was popular at the time of the growth of mass education early in the twentieth century, and which is often associated with methods of training and conditioning. It has been shown that animals can be conditioned into displaying particular desired traits, or features of behaviour. Humans can also sometimes be conditioned, but such methods do not necessarily lead to understanding, creativity, or the ability to solve problems. Stimulus–response based activities, such as rapid-fire mental tests or 'flash' cards, are sometimes appropriate, but basically they are intended only to improve memory and recall, not to promote original thinking. However, if we were to decide we wished to produce a generation of children conditioned to accept and skilled in unintelligible routines, and not required to think, then sole reliance on particular behaviourist methods might be appropriate. Therefore the nature of the kind of learning outcomes we wish to promote should enter into considerations of the choice of teaching method. If our aim is to assist children to develop into thinking, problem-solving adults, we need to use teaching methods in school which are most likely to foster such competencies. The climate in which this book is being written is one of international competition and league tables, and the pressure to 'create wealth', which surely requires ingenuity and innovation rather than facility with routine procedures. In order that mathematics teaching supports both personal intellectual development and the interests of economic progress, it has been claimed that 'problem solving must be the focus of school mathematics' (NCTM, 1980, 1989, p. 6). Methods of teaching practised by inspired teachers of two thousand or more years ago, such as Socrates and Jesus, reveal that methods other than exposition are most certainly available to us, and that simple transmission has only relatively recently become 'traditional'.

UNDERSTANDING MATHEMATICS

Considerable emphasis in learning mathematics in recent years has been placed on the desirability of understanding, rather than on being able to repeat remembered routines and demonstrate particular basic skills. How we learn is such a complex issue, however, that it is possible neither to separate understanding from memory nor to explain what is meant by 'understanding mathematics'. In relation to memory, the more readily one remembers the easier it is to think, because there is little delay caused by searching for what can be likened to some missing piece of a jigsaw puzzle, and because less effort is required in pulling essential information to the forefront of the mind. Pupils whom teachers regard as being particularly intelligent usually have swift and reliable retrieval systems, in that they recall things quickly and accurately.

Is this all there is to understanding? This is a debatable point, but it seems likely that a good memory is only a part of what is involved in understanding. Some years ago, Skemp (1976) introduced us to the notion of two kinds of understanding mathematics, namely instrumental and relational. When children are learning many of the procedures of mathematics, such as adding two fractions or multiplying two three-digit numbers, it is possible for them to think they understand because they can nearly always achieve the correct answer. Often, children will claim they understand when all they know is what to do, which is certainly dependent on memory. What they do not necessarily know is why they do it, or, in other words, why what they do produces the correct sum of the two fractions, or the correct product of the two three-digit numbers. This latter kind of understanding seems much less a feature of memory than knowing what to do. To know what to do is to have learned 'instrumentally'; to understand why is to have learned 'relationally'.

Teachers are likely to desire that children will learn mathematics relationally, but must at the same time suspect that this often does not happen. It is even tempting to wonder whether it is possible for many children ever truly to understand why the algorithm (a mathematical technical term which we shall equate with 'procedure') for long multiplication works, let alone long division. This leads to the question as to what is the effect on the children, over a long period of time, of learning, practising and revising algorithms which it seems may not be relationally understood. Some children will subsequently achieve relational understanding, and in the meantime will not be too adversely affected by having to suspend full understanding. Many other children, however, will not be able to learn algorithms in an instrumental way. For such children it could easily be the case, first, that mathematics appears to be composed largely of procedures which they just have to do without much hope of understanding why they work; secondly, that mathematics is a mysterious or even magical subject which produces correct answers even though hardly anyone knows why; and thirdly, that mathematics is not for them because they cannot properly understand what it is all about. One reason why many adults never mastered much mathematics could be that we have too often been taught it as if it consists of procedures which cannot be understood, or which it is not necessary to understand. In terms of relational understanding, methods like exposition and practice are generally unsatisfactory, because they encourage this algorithmic approach to learning, and we need to consider alternative methods such as the others listed by Cockcroft.

One reason learners encounter difficulties in understanding mathematics is its abstract nature. As with 'understanding', we again run into problems of definition, for according to some dictionaries 'abstract' means 'not concrete', and 'concrete' means 'not abstract'. 'Concrete' suggests an object which we can literally get hold of, so the metaphorical equivalent of 'getting hold of' an idea is probably also appropriate. Some ideas in mathematics seem easier to 'get hold of' than others. Yet it is all relative. Even the numbers we count with are abstractions, derived from objects which we can see, hold, place together, separate, combine, share out and generally rearrange and manipulate. But most children eventually become at ease with handling the 'natural numbers' (the numbers we count with), almost as if these numbers have become concrete entities. Indeed, when children are trying to learn about fractions, decimals, integers, ratios and percentages, and also take on board concepts like place value, proportionality, similarity and function, the basic counting numbers seem like old friends, familiar and comfortable to be with. Learning mathematics is like learning to cope with abstractions built upon abstractions – yesterday's abstractions can become the relatively concrete ideas on which we attempt to build today's newer abstractions.

A major problem for many children occurs when they move into the realms of what is usually described as algebra, which, although undoubtedly abstract, is basically built only on the familiar counting numbers, and in that sense does not seem to be all that far removed from them. Algebra is an important part of mathematics, in that it not only provides a shorthand notation, but also, for mathematicians, makes thinking and reasoning easier, facilitates generalization, and provides a symbolism with which to conduct logical arguments. To some, algebra is the very essence of mathematics. But most people do not see algebra that way: they see it as making their thinking and reasoning more difficult, they see it as a major obstacle to learning. According to Cockcroft (1982, p. 60), 'algebra is a source of considerable confusion and negative attitudes among pupils'. The fact that many children cannot understand much of the algebra we teach leads them to dislike and ultimately reject it, and thence possibly even the whole subject of mathematics. The particular kinds of abstractions involved in algebra often present a very real learning difficulty. In order to facilitate learning, we need to adopt teaching methods which assist children to move smoothly from the relatively concrete to the relatively abstract at all stages of their mathematical education. This applies at the very earliest stages of learning, when too early an introduction to formal written arithmetic can cause fatal problems – in the sense that the children never recover – just as much as to later stages when simple linear equations are introduced, using the symbolism of algebra.

The issue which immediately follows is that of 'readiness'. Must we wait until children are ready before it makes sense to try to teach, say, algebra? Here again, there is no simple answer. Certain significant theories concerned with learning have been based on the view that there is a sense in which it is necessary for cognition to mature before letters can be used instead of numbers, and the concept of a variable can be introduced. The idea that it might be necessary to wait until a pupil is ready is one which many teachers find meaningful. After all, at one particular age a child might be unable to comprehend place value or the use of letters for numbers, but a few years later there might be no problem. In the intervening period of time, however, the child has learned a great deal from a wide variety of environments. There is no way of being certain of the relative contributions of experience and greater maturity of the intellect, and, in any case, experience presumably contributes to increased maturity. The one thing we can be sure of is that if at a given time we are unable to teach an idea to a particular child we can always try again later. And if we do not want to alienate children from mathematics, perhaps the best course of action is not to push too hard against a brick wall. Another possibility is to look for alternative teaching methods. It was Bruner (1960, p. 33) who claimed that 'any subject can be taught effectively in some intellectually honest form to any child at any stage of development'. This does not mean that algebra can be taught to five-year-olds. It must seemingly be interpreted as suggesting that there are some valuable experiences which five-year-olds can encounter which will start them on the road towards algebra. Always there are mathematical experiences which will move a child on, in terms of relational understanding and deeper cognitive development. Progression is important, and should always be possible, it is just that any particular child might not be able to progress through a curriculum as quickly as teachers might hope.

Learning mathematics requires pupils to understand concepts, and the difficulty is that concepts cannot be learned without intellectual effort, frequently over a considerable period of time. Concepts are abstract ideas such as 'equality', 'work' and 'quality of life'. In mathematics, some examples of concepts are 'real number', 'similarity' and 'function'. Mathematics incorporates a multiplicity of concepts, and some might be thought of as more abstract than others, thus suggesting a kind of hierarchy. The concept of 'triangle' is relatively concrete in that it is possible to draw, make, hold and generally manipulate triangles. But we must not overlook the fact that there are many conceptually difficult ideas involved in a deeper study of triangles, for example:

- that triangles are polygons with exactly three angles;
- that angles are measures of 'turn';
- that polygons are plane shapes with straight edges (sides);
- that a plane is a flat surface in space;
- that there is potentially an infinity of planes in space;
- that there is potentially an infinity of lines in any plane;
- that an edge is a straight line;
- that a line which is straight forms the shortest distance between its two end points;
- that not all lines need have end points;
- that a point is a location without size;

and so on. It is clear that mathematics contains a multiplicity of ideas, many of which are not easy, and to burden young children with such a large number of concepts would be ridiculous, indeed impossible. The concept of 'triangle' is learned or developed by seeing, holding and making triangles, by approaching the idea through examples and in as concrete a way as possible. It is possible to provide a definition for 'triangle', but a definition is unlikely to help most children at the beginning of learning about triangles. We learn from examples and counter-examples, though a definition might be useful as a summary later, particularly if it is composed by the children themselves. We also learn from using all our senses, in mathematics most particularly sight, hearing and touch, touch being very important in learning concepts such as 'triangle'.

A less tangible concept, which appears to present learners with greater difficulty than 'triangle', is that of 'fraction'. A common conventional concrete approach to learning about fractions is to think of parts of wholes, like cakes or bars of choco-late. It is wise, however, also to base the idea on discrete countable objects like sweets (in a packet, say), or a class of children, as well as on objects like bars of chocolate which can be thought of as requiring the dissection of a continuous whole. This not only helps to broaden or extend the understanding of 'fraction', but provides examples on which a subsequent discussion of discreteness and continuity, both very important mathematical concepts, might take place. However, this concrete approach to fraction does not provide the whole story. To mathematicians, fractions are rational numbers, not parts of wholes. They are not only ratios of two natural numbers (in fact, rational numbers should perhaps be thought of as ratios of integers), they are numbers themselves. Thus it is impor-tant also to link the concrete approach to fractions with two independent ideas, first of ratio (hence the term 'rational') and secondly of relative positions on a number line (see Figure 2.1).

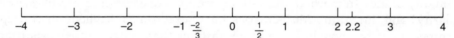

Figure 2.1   Fractions on a number line

Natural (counting) numbers (1, 2, 3, 4, 5, . . .) are usually accepted by children, but other number sets, including the integers (. . . -5, -4, -3, -2, -1, 0, 1, 2, 3, 4, 5, . . .), the rational numbers and the real numbers (both of which cannot sensibly be listed), can probably never be quite as concrete as triangles, and are not as easy for children as many well-educated adults might think. Subsequent curriculum topics such as ratio and proportion, or the 'nesting' of the number sets, are even more complicated (see Figure 2.2).

Ratio leads naturally to the idea of 'similarity', a concept which is difficult to understand because there is not only mathematics to cope with, but also a language problem, namely the fact that 'similar' carries a more specialized meaning in mathematics than it does in the everyday world of children. Concepts like 'rational number' and 'similarity' can be internalized only over a relatively long period of time. They cannot normally be learned instantly or by definition, and may even never be properly learned at all. A more extensive discussion of mathematics learning is provided by Orton (1992).

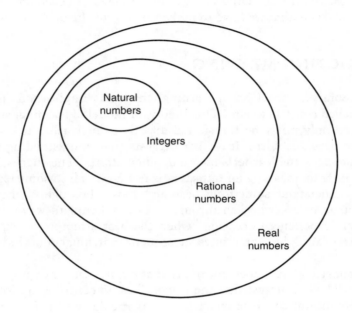

Figure 2.2   Nested sets of numbers

Another issue which faces teachers of mathematics is that society might *demand* that we teach routines like the addition of fractions and long multiplication, even though relational understanding might not be achievable by many of the children. Although it is always wise to strive for relational understanding, it is clear that as a last resort we sometimes feel compelled to encourage children just to remember a rule or algorithm. Many children buckle under the weight of a constantly growing burden of such rules and procedures whose purposes are not understood. Other children, however, seem able to accept procedures for what they are, and they become routine, and the children can then move on to learning the next one. For some such children, light dawns later, and they survive, but unfortunately this dawn of understanding occurs for far too few. There are many adults who, as children, learned mathematics in this way, without it particularly worrying them that they did not know why algorithms worked, and some of these people work with mathematics in adult life, having perhaps steadily been able to derive meaning for the procedures. Yet there are many more adults for whom mathematics is an unpleasant memory largely because it was, for them, devoid of meaning.

Here we have a difficulty. We can never be sure which children will be content and satisfied with deferring understanding until some later date. The suspicion is that it is only the more able children who can accept the suspension of under-standing in this way, and that the majority of children cannot. (This raises many questions about 'ability', a widely used word without a unanimously agreed meaning, which we do not have space to discuss here.) It is particularly for the majority of children, then, that we need to strive for relational understanding of what we teach. In an ideal world, teachers would be given the opportunity to construct a child-friendly curriculum which would enable them to teach what was learnable by each particular child at each age or stage of development. Unfortunately, it often seems that we have to make the best of the one laid down for us.

CONSTRUCTING MEANING

One contemporary perspective on learning which has implications for teaching methods is called constructivism. This theory, or model of learning, suggests that knowledge can only rarely be transferred directly from teacher to learner in an immediately 'digestible' form. It further suggests that understanding usually has to be constructed by the learner's own individual efforts. The suggestion that we all need to construct meaning for ourselves is not new, and to some extent is self-evident. We understand something best and most thoroughly when we have worked it out or at least checked through it in a meaningful way ourselves. We have all surely experienced occasions when the best attempts to explain something to us have failed, but sometimes subsequent individual effort has resulted in mastery.

However, constructivism does not imply that pupils can make progress only on their own, nor does it suggest that the teacher has no contribution to make. Nor does an acceptance of constructivism lead to particular well-defined methods of educating children. In the past, dissatisfaction with widely used teaching methods has led to views about the need for children to be actively involved in their own learning, and these have then led to the commendation of practical work, structural apparatus, 'discovery' learning, and most recently to investigative methods. Constructivism does not commend any particular methods, but nor does it rule them out. In fact, the methods listed above would be likely to feature heavily within a constructivist approach to teaching. Most importantly, the teacher should not sit back and wait for light to dawn on each child, but rather has a vital role to play in arranging the environment so that learning with understanding is fostered for each child. The teacher can design tasks, assignments, problems, projects and other activities which will stimulate thinking and mental activity, and which are thus likely to lead to the construction of meaning. The teacher might also sometimes guide and prompt in a fairly direct manner, but is unlikely to engage in much 'telling'. The role of the teacher was summed up by Richards (1991, p. 38) in the words, 'Students will not become active learners by accident, but *by design*.' The acceptance of constructivist views suggests a particular role for the teacher in the classroom which is certainly different from, and which may be much more demanding than, exposition and delivery.

Constructivism exists in the minds of theorists in a number of forms, and only the basis has been described above. Radical constructivists suggest that, because knowledge and meaning are constructed by each individual, there can therefore be no certain and objective knowledge about the world. The problem is that we might all draw different conclusions from our environment, and the difficulty is compounded if we try to compare our conclusions, because language is notoriously imprecise and is itself open to idiosyncratic meaning and use.

However, another version of constructivism emphasizes the value of social interaction and communication, in other words children working together in pairs or in small groups. The value of discussion as a way of enhancing learning has been accepted by educationists for many years. According to the DES (1985, p. 39), 'The quality of pupils' mathematical thinking as well as their ability to express themselves are considerably enhanced by discussion.' Discussion can assist learning at any level, because the very articulation of thoughts lays them open for inspection, criticism and amendment, thus, one hopes leading to clarification and a coming together of understandings. An important aspect of the teacher's role in such a situation is to set up the groups appropriately, so that meaningful exchange can take place. One very practical point is clearly not to make the groups too big, otherwise some children play little part and learn less than they might. The construction of meaning takes place in a group situation by means of negotiation between group members, with discussion eventually leading to mutual acceptance. What is agreed is what is socially accepted, and thus if it is true that there is no such thing as certain knowledge, at least discussion is more likely to lead to what society expects us to believe. Here we can begin to appreciate the relevance of the suggestion in the Cockcroft Report (Cockcroft, 1982) that we include 'discussion between . . . pupils' as one of our teaching methods. A more detailed consideration of constructivism and the role of discussion in mathematics teaching, and problems associated with its use, is to be found in Orton, A. (1994).

Cockcroft (1982) also suggested there was a need to incorporate 'appropriate practical work'. Although constructivism does not advocate the use of practical work in which children handle and manipulate concrete materials, there is a long history to the commending of the use of 'manipulatives', from the time of Froebel, through Stern, Cuisenaire and Dienes, to today, and there are many who would expect a constructivist classroom to contain a wealth of manipulative materials. Cockcroft stressed that 'it is too often assumed that the need for practical activity ceases at the secondary stage but this is not the case' (1982, p. 72). Moreover, there are many who believe that practical activity does not form a large enough component of mathematics lessons in the primary school, particularly with the older children. With very young children, practical activity is all that is possible, and this is widely accepted, but it is the move away from the concrete to the more abstract which is often hurried, and is another reason why some children fall by the wayside on the road of mathematics learning. The main problem is the pressure to progress through the curriculum, including the expectations of parents, unfortunately at the expense of understanding. Another difficulty is that it often seems as if, no matter how much time is spent on practical work, some children are never ready to abstract, and that if the more abstract equivalent to the concrete activity is introduced, children do not see the link (Hart, 1989). It is in bridging this gap between concrete and abstract that teachers have an important role to play, and discussion between teacher and pupils would seem to be essential. Practical work is a valuable vehicle for assisting pupils to construct meaning, but success cannot be guaranteed, and the teacher must also expect to have to work hard in trying to move the pupils forward in their mathematical thinking.

Cockcroft (1982) also advocated the use of 'problem solving' and 'investigational work', and this concurs with the recommendations of the NCTM (1980, 1989), referred to earlier. It is important to emphasize that 'problem solving' is not intended to refer to routine questions arranged in 'exercises', such as are scattered throughout mathematics textbooks. Rather, it refers to the use of novel problems which require children to draw upon previously acquired knowledge and expertise in an intelligent rather than random or routine way. Some such problems should be related to 'everyday' or 'real-life' situations, according to Cockcroft, and others need not be. The intentions and advantages of solving problems which contain an element of novelty should be obvious. In the first place, this is what being a mathematician is all about, and in the second place, the children are clearly forced to construct or develop meaning in order to make progress. Problems force children to think, to recall and use relevant prior knowledge, and to be creative in a modest way.

Again, the role of the teacher is a difficult one. The whole point is that it must be the children who do the solving, and any direct intervention possesses the danger that it will drift into exposition. Yet the teacher does have to be ready to step in when an impasse has been reached, or when frustration is observed, and then provide a hint or, better still, ask a question which will promote construction. Once again, however, we see that this particular teaching method can be related to theoretical views on how learning is most likely to occur. Investigational work falls into the same category, in the sense that it demands originality or creativity from the pupil. The issue of investigative approaches and problem solving will be dealt with in more detail later in this book.

One approach to the learning of mathematics which generally appeals to children and which often encourages construction is via pattern. This can be taken to include not only patterns of sets of numbers and shapes, but also patterns within shapes and other mathematical structures, even the structure of large areas of the entire subject. There is something within most of us which not only attracts us to patterns, but also enables us to gain insights from them. Sometimes, too much time is spent in mathematics lessons on colouring patterns, like tessellations for example, but if they provide us with both motivation and cognitive support, then it is essential we take full advantage of the approach.

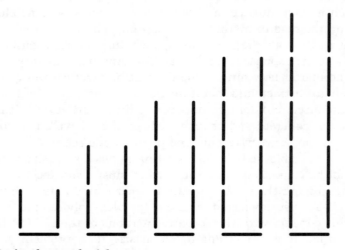

Figure 2.3   A simple matchstick pattern

Earlier, the introduction to algebra was used as an example of the difficulty of moving from the concrete to the more abstract. If it is required that children should construct their own understanding or meaning, then environments which encourage this are needed, and in recent years the use of number and other patterns has been advocated as just such an environment. The pattern of shapes constructed from matchsticks in Figure 2.3 leads to the number pattern 3, 5, 7, . . . Children can be asked to assume that the shapes constitute the first three of a sequence which can be extended to the fourth, fifth, tenth, hundredth, and even the nth shape. The relationship between the 'shape number' and the number of matchsticks requires an appreciation of 'twice the shape number plus one', which many consider is a step towards an algebraic representation. The introduction of the idea of an 'nth' term demands the final step, namely a formula like $M = 2n + 1$. At a much earlier stage, very young children can be encouraged to generate patterns with beads, counters, pegs, and the like. Many such tasks are possible, and this issue will be followed up in Chapter 9. Indeed, reference to pattern as an approach to learning mathematics will be used whenever appropriate in this book. In the context of the present chapter, pattern might well sometimes offer a means of generating the construction of mathematics by the children.

CHAPTER 3

Problems and investigations

INTRODUCTION

A discussion of problems and investigations is often difficult to conduct as both concepts are seldom clearly defined, or understood. Many authors deliberately assume that they are interchangeable; others suggest that there are differences, but then fail to explain what these are. A third group acknowledge that there are differences, but argue that in the classroom such differences are unimportant. In this chapter an attempt is made to provide a working definition of both mathematics problems and investigations and to show that it is important that students as well as teachers should be conscious of the care with which the words should be used in a classroom.

PROBLEMS AND THE CURRICULUM

Over the last two decades mathematics educators have stressed the importance of the role of problems in the mathematics curriculum and in the development of students' processes and skills in solving problems. Johnson and Rising (1967, p. 104) claim that 'learning to solve problems is the most significant learning in *every* mathematics class'. They give the following reasons:

1   It is a process whereby we *learn new concepts*.
2   Problems may be a meaningful way to *practise computational skills*.
3   By solving problems we learn to *transfer concepts and skills* to new situations.
4   Problem solving is a means of *stimulating intellectual curiosity*.
5   *New knowledge is discovered* through problem solving.

These are not exaggerated claims, but ones based upon many years of teaching mathematics. However, they lack the support of hard, conclusive research. Yet, as is stated in Chapter 1, mathematics is increasingly seen as a subject that is able to develop in children the ability to solve problems. There is little evidence to indicate that the knowledge and skills which children develop when learning to use and apply mathematics transfer to the solving of problems in other areas of the curriculum. Nor is it known whether the exploring of mathematics problems in the curriculum is of help to school-leavers when they start new careers in industry or commerce. We must be prepared to challenge any assumption that mathematics problem solving is transferable beyond the boundaries of mathematics.

This is not, however, to suggest that the study of mathematics problems has no place in the development of students. A study of problems in mathematics may make a significant contribution to the development of the whole child equal to that of other subject areas that make the same claim. Despite the lack of evidence about the value of problems most teachers and educators remain convinced by their experience in the classroom that solving problems merits inclusion in the mathematics curriculum. Way back in 1967 Professor Geoffrey Matthews in *I Do, and I Understand* wrote 'The solution of genuine problems and the judgement-making-involved are integral parts of living' (Nuffield Mathematics Project, 1967a, p. 13), the implication being that this is a justification for problems being in the mathematics curriculum. The belief is expressed with conviction that, although the content of a mathematics curriculum is essential, the focus on problem-solving shifts the weight from the acquisition of knowledge and skills to using and applying them. However, some advocates of a problem-solving curriculum also claim that mathematics cannot be meaningfully taught, as it is the creation of each individual interacting with their environment. This is reminiscent of a constructivist perspective on learning (see Chapter 2).

A SHORT HISTORY OF PROBLEMS IN THE CURRICULUM

Since the early days of formal education problems of a certain kind have been part of the mathematics curriculum. In general, they have taken the form of problems set in words where the mathematics is contextualized in pseudo-real situations. The justification for their inclusion is that they enable students to practise the knowledge and skills which have been previously taught in a 'real' context. As Matthews (Nuffield Mathematics Project, 1967a, p. 13) succinctly writes, 'a problem . . . meant a piece of mechanical arithmetic disguised by the use of words'. The teaching of this type of problem is very standard, with a teacher showing how to solve an exemplar problem on a limited aspect of mathematics followed by students practising the techniques learned on similar problems. The aim of this approach is to establish techniques so that on meeting a problem having the same mathematical structure in an examination the appropriate knowledge and skills are immediately recalled and applied successfully. Many readers of this book will no doubt have experienced this approach as students in school and in higher education. Some will also recognize it as a method of teaching they use with their children. A quick flick through today's mathematics textbooks at primary and secondary levels shows that problems of this nature are still the staple diet of many students. Problems may now be accompanied by attractive illustrations and diagrams, but they are basically the same kind of 'word problems' which have been in the mathematics curriculum for many years. There is still a role for such problems in a student's experience; they should not be dispensed with to be replaced by a passing 'bandwagon' fad of the late twentieth century.

For over sixty years problems and problem-solving have been important areas of research for educators and psychologists. Brownell (1942) and Polya (1957) are two of the better-known researchers in this area. However, it was not until the 1960s that a new and different perspective on mathematics problems emerged. In 1967 the Association of Teachers in Colleges and Departments of Education (ATCDE) asserted that mathematics problems exist in their own right, being in nature different from general problems which had been studied in the past. They have intrinsic interest and, in general, a problem-solver should have no immediately obvious techniques available for their solution. The ATCDE further claimed that this type of problem 'is the essential characteristic of real mathematical activity'. This attempt to change the focus of the mathematics curriculum toward a newer and very different view of what problems are and their role in learning mathematics was supported by the Association of Teachers of Mathematics (1969). This body promoted the view that a problem, perceived as a situation to explore, is a more valuable mathematical task than one involving a reproduction of a ready-packaged method applied to recognizable problems set in words. The objective was to foster in students a belief that they made mathematics their own through exploration. The NCTM (1980, p. 1) in *An Agenda for Action* advocated that 'problem solving should be the focus of school mathematics in the 1980s'. In 1982 the Cockcroft Report gave the seal of approval to the place of the more modern idea of problems in the mathematics curriculum asserting that 'mathematics teaching at all levels should include opportunities for problem solving, including the application of mathematics to everyday situations' (1982, p. 71). In the last fifteen years a number of official reports have lent authority to the inclusion of problems in their newer form in the mathematical experiences of students.

After a slow beginning the contemporary approach to mathematics problems has received a measure of approval from most teachers. The newer types of problems are now appearing in textbooks both at the primary level and at the secondary stage where extended coursework at GCSE has promoted their inclusion. More and more children are experiencing meaningful situations as the settings of problems in mathematics. Yet, despite overwhelming pressure from august educational associations and numerous recommendations to make problem-solving an important goal in mathematics teaching the translation of this aim into practice has yet to bear noticeable fruition. In many primary schools where teachers are not required to prepare children for external examinations there is little evidence that word problems are complemented with situation problems created and explored by children for their own sake.

WHAT ARE PROBLEMS IN MATHEMATICS?

Psychologists and educators do not have a readily agreed working definition of a problem. Mathematics educators have, also, yet to arrive at an acceptable description of what a mathematics problem is to which all teachers could and would adhere. Johnson and Rising (1967, p. 104) suggested that solving problems 'is finding an appropriate response to a situation which is unique and novel to the problem solver'. The ATCDE (1967) talk of open problems, implying that there were categories of problems, without saying what a problem was.

There are differing perspectives to consider when searching for an adequate and workable definition of a mathematics problem. Not all children in a class may view what is taught as a problem. Those who have little understanding of a situation will view any mathematical idea arising from an activity associated with the situation as a problem. Children who have already met a situation before and become reasonably familiar with the different aspects of the mathematics of the activity will view their work as a repetitive exercise. Thus what is a mathematics problem for one learner may be an exercise for another.

Consider the familiar addition 6 + 3. A student experienced in using numbers and performing operations on them is likely to recall the answer 9 without apparent effort. In no way is it possible to claim that this student sees the addition as a problem. However, to a child who is at the early stages of developing an understanding of the concept of number and its associated symbols this is indeed a problem. What are the differences between the two learners which enable us to decide that 6 + 3 is a problem for the latter child, but not so for the former? The student who readily answers is likely to have previously experienced similar additions. Not only is the situation familiar, but a goal is readily recognized and is quickly achievable, either because an answer is recallable or a technique is available which enables the answer or goal to be quickly calculated. Indeed, such a student would never consider 6 + 3 to be a problem. In contrast the student faced with 6 + 3 and having no immediately obvious way of determining the solution is faced with a problem and assumes the existence of an achievable goal. Thus the challenge inherent in achieving the answer must also be part of the problem. Many methods, processes and strategies are likely to be open to this student to arrive at a goal. A relatively sophisticated five-year-old may use fingers to count 6 and then count on 3, to arrive at 9. A less advanced student may need to ask what the '+' sign means before exploring the combining of the numbers, perhaps by modelling the mathematics using objects, recounting the total set after a constructing a set of 6 and a set of 3. From this one example it is possible to list the key aspects of mathematics problems. A mathematics problem can be said to be a situation in which an individual student:

(a) recognizes or believes that there exists a mathematical goal to be achieved, usually an answer of some kind;
(b) accepts the challenge to perform some mathematical task in order to reach the goal;
(c) has no readily known or recallable mathematical procedure available to enable the goal to be attained directly.

Here are some mathematical situations or activities which occur in a classroom. Without attempting to explore them try to decide which ones, for *you*, are problems in that all three of the above criteria are satisfied.

Activity 1   *How many pairs of numbers are there which add up to 475?*
Activity 2   *How many squares are there in this grid?*

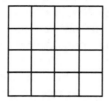

Activity 3   *How much does it cost to travel from London to Paris?*
Activity 4   *How many 1p, 2p and 5p coins are needed to make 20p?*
Activity 5   *Explore square numbers.*

The first two activities have recognizable and unique goals. However, many teachers will know children who conveniently change the goal so that it no longer matches the original intention. Would you feel motivated to answer the questions? The desire to answer a question is very personal for children when teachers present them with a mathematics activity. Unless the two situations, or something like them, have been experienced before it is very unlikely that you could immediately decide on a method or strategy for arriving at an answer. With a little thought and perhaps more time you may be able to formulate a procedure which you could try out. Perhaps by changing the goal in both activities you may find more efficient procedures which not only arrive at the answer to the original question but generalize to other problems.

The cost of travelling from London to Paris in the third activity has an obvious goal, the cost, but there are many possible answers as the cost is dependent on the mode of transport and the route taken. Thus goal and answer are not necessarily the same. Unfortunately, we too frequently lead children to believe that there is always a unique answer to a problem. The procedure for the attainment of the goal may be available to you if you have made trips of this kind before and can apply your previous experience to this journey.

The fourth activity has the goal stated in the question. It is also one which you may have experienced when buying an item for 20p. In such circumstances you are motivated by the desire to purchase the item. Experiencing this situation in a classroom, however, is very different and may not be considered a challenge by some children. Once again many solutions are possible, but this may not be immediately recognized by children, particularly those who are satisfied to have achieved any answer. Children should compare answers so that they recognize that more than one is possible. The interesting aspect of this question is the strategies which students use to ensure that all possible combinations of 1p, 2p and 5p coins which make 20p are found. The search for an efficient and effective strategy often provides the motivation to attain the goal.

The fifth and final activity may be said to have a general goal, namely the exploration itself. However, no answer is apparent as no question has been set. How the exploration is conducted gives rise to specific goals and subsequent answers. Thus both goals and answers only arise as a consequence of the exploration. A procedure is not obvious at the start unless the activity is familiar in that it has been done on other kinds of numbers. Many students find such a vague and open request to explore daunting and inaccessible as so many decisions have to be made before goals and procedures arise. Very few mathematics educators would classify explorations of this kind as problems as they fail to satisfy at least two of the criteria described above. This activity is more likely to be known as an investigation of which more will be said in the latter part of this chapter.

CATEGORIZING PROBLEMS IN THE CURRICULUM

There are three major categories of problem in which teachers and students operate. *Routine problems* use knowledge and techniques already acquired by a student in a narrow and synthetic context. In the classroom routine problems occur as sanitized word problems which are often fabricated by textbook authors. Here is one such routine problem:

'*How many more than 286 is 637?*'

This is an example of many similar problems which seek to translate a symbolic mathematical statement, in this case 637 − 286, into one which uses words as well as symbols. There is a sound justification for some of these problems as they can occur in situations which students may meet in their everyday lives. When these kinds of questions arise students are expected to understand the linguistic complexities of the word problem and either model it in mathematical symbols or respond with the appropriate mathematical operation. Here is a second example of a word problem:

A postman has ninety-four letters in his bag.
Twenty-five of them are first-class. How many are second-class?'

This type of routine problem is often called a story problem in words. As such it comes in the category of word problems set in a 'real' context. Before students commence the process of solving problems of this kind they require some understanding of the contexts in which the problem is set. A student who is unaware that letters are referred to as first- or second-class because of the stamp that is on them cannot make a start on the solution of the problem. A student may respond to the problem by asking for more information about how many different classes of letter a postman carries in his bag. This is a legitimate request. Many story problems make many such assumptions about pupils' experience and prior knowledge.

You will recognize that when students are given word problems the purpose is limited to the cognitive consolidation of facts and techniques recently taught. Practice of limited routine problems is not an appropriate method for the development of new knowledge and its contribution to mathematics learning is minimal. Indeed you will have noticed that such problems fail to satisfy the third criterion which required that a problem-solver does not have a readily available procedure for solving the problem. There are those who would challenge the notion that routine problems are problems. But there is little doubt that teachers ask their students to work on routine problems and that they do refer to them as problems.

Environmental problems, often called 'real-life' or 'real-world' problems, are set in contexts which represent the real or practical world, or as close a match to the real world as possible. The organization of an inter-school sports competition would be an example of an environmental problem. The mathematics in an environmental problem is used to find a solution. Once this is achieved there is no need to draw out the mathematics within it. In such instances mathematics might be purely a tool and not an end in itself. Problem-solving in the real world can have purpose in mathematics lessons only if the contexts in which the problems are embedded have familiarity or make sense to the students. There is little point in asking students to solve problems chosen from the environment in which they live if they are unable to relate to the problems and have no desire to solve them. It is the extent to which real or practical problems presuppose some familiarity with the context and content background of the physical world that makes it difficult to incorporate them into the mathematics curriculum. A fourteen-year-old girl was given the problem of working out the possible half-time scores in a football match when the final score was 2–1. She responded by claiming she could not do the problem because she did not know anything about football. How often do teachers 'create' environmental problems which they feel are interesting and challenging only to find them rejected by pupils who are unable to relate to them? Johnson and Rising (1967, p. 109) asserted that 'for most students it is *not* necessary to take problems from their immediate environment. Often, students are less interested in grocery bills than in cannibals and missionaries, less interested in volumes of oil tanks than in walks through Koenigsburg.' (The references to cannibals and missionaries and to walks through Koenigsburg are to two widely used non-environmental school mathematics problems.)

Environmental problems require students to apply informal as well as formal knowledge to their solution. Informal knowledge is developed from direct experience, seldom from the formally constructed knowledge of the mathematics classroom. Students also bring to such problems individual heuristics and combine them with routine procedures in creative and unexpected ways. The interaction between these two aspects of learning provides the most favourable conditions for the development of a student's mathematics. Frequently mathematics arises which is new to the student. 'Old' knowledge is consequently reorganized and restructured into an expanded and aggregated body of 'new' knowledge.

It would appear that environmental problems should be an important and essential feature of the mathematics curriculum. Why then are they so infrequently found at work in the classroom? One major reason lies in the adherence of teachers to 'textbook' mathematics. Real problems which match the interests of students cannot be written down in a textbook for all to use successfully. The moment that a real problem is set in print it becomes a word problem with its subsequent difficulties and artificiality, losing its sense of reality. Environmental problems should arise from a student's own experience which may not be that of other students in the same class. Problems set in the real world of students require the use of mathematics procedures and thinking processes that may be very different from those learned at school. For example the solutions to many real problems do not demand the accuracy of calculation which may be deemed appropriate in a mathematics lesson. Approximation and estimation can be the most suitable skills for a task when solving a real problem in contrast to the exact algorithmic procedures which are taught in the classroom. A great deal of the mathematics which appears in the curricula of countries throughout the world is inappropriate and irrelevant to the solving of environmental problems. Often there is a mismatch between a national mathematics curriculum and its use in solving real problems. When students solve real problems a teacher loses control of the mathematics which they use and develop. On the one hand students use mathematics not taught in the classroom, and on the other hand they develop new mathematics which seldom appears again in a mathematics lesson.

Process problems are set in a mathematics context in contrast to real problems. This type of problem concentrates on the mathematics itself and on the mathematical thinking processes for arriving at the solution. In many such problems the answer may not be important, taking second place to the method of getting to the answer. It is this latter aspect of process problems that forms the appeal and the challenge. Activity 1, *'How many pairs of numbers are there which add up to 475?'*, is an excellent example of a process problem. The goal is obvious, yet several methods are available to the solver to arrive at an answer. Children who reflect upon the processes and strategies (a sequencing of different processes) they use to solve a process problem contribute to the development of their ability to solve other problems, similar yet unrelated. Some of the mathematical and thinking processes used to solve process problems become apparent when Activity 1 is solved. Below we examine two possible methods which students adopt to find the number of pairs which add to 475.

The first method begins with a random choice of pairs of numbers such as 200 and 275, 122 and 353, and 400 and 75. Students who start with this approach very quickly recognize the need for a more systematic strategy. An ordered listing of pairs takes place beginning with smaller numbers:

> 1 and 474
> 2 and 473
> 3 and 472 etc.

After some thought, discussion and the search for patterns it is noticed that the first number in the pairs in the ordered listing can be used as a counter of the number of pairs discovered. When the list nears its end the last pairs are

> 472 and 3
> 473 and 2
> 474 and 1

As the first number of the last pair is 474 the number of pairs which add up to 475 is 474.

A second approach recognizes that to work with a large number like 475 is wasteful of time and energy. If only the number was much smaller, say 5. In that case it is easy to write down the required pairs: 1 and 4, 2 and 3, 3 and 2, 4 and 1, a total of 4 pairs. What if the number had been 6, 7, 8, . . .? A pattern becomes apparent, particularly if the information about the number and the total number of pairs is listed in an ordered way; that is, in a table (see Table 3.1).

Table 3.1

	The number	Total number of pairs
	5	4
	6	5
	7	6
	8	7
	.	.
	.	.
	.	.

The information in the table suggests that the total number of pairs is one less than the number.

We must remember that the observed pattern has been confirmed only for the numbers 5, 6, 7 and 8. You may feel strongly, or even that there is a certainty, that the pattern will continue. However, such feelings are not proof of its continuation as far as 475. Inductive reasoning as opposed to deductive reasoning is a thinking process which is used frequently in the solving of process problems. Although mathematics teachers are desirous that a solution arrived at by inductive reasoning be followed by deductive proving of the conclusions, students seldom see the point in confirming what they already know.

PROCESSES AND SOLVING PROBLEMS

There is much confusion in the mathematics education literature on problem-solving and processes. The two methods used to solve Activity 1 illustrate a number of processes at work. In both cases information is collected that contributes in some way to the solution of the problem. The collecting of data is a process which is used in many ways to assist in solving process problems. The mere collecting of information is not in itself sufficient. The data have to be 'operated' on to make sense of what they are telling us. Both methods ordered the data. The collecting and ordering of data are *operational* processes. The ordering took place as a list in the first approach and as a tabulation in the second. Listing and tabulating are *recording* processes. The search and observation of patterns took place in both methods. Pattern-searching is an important *mathematical* process (see Frobisher, 1994), which itself depends upon analysis of and reflection upon data, noticing similarities and differences. Other examples of mathematical processes 'unique' to the subject are guessing, predicting, generalizing and proving. Analysis and reflection are *reasoning* processes, as are clarifying and understanding. The description of the methods is a *communication* process. In a classroom it is highly likely that children will be involved in other processes of communication, such as explaining, talking, agreeing and questioning. As processes are used more and more often in different circumstances and with a variety of problems and investigations they gradually take on the nature of skills. A process becomes a skill for children when it is appropriately selected to match the needs of the situation and can be applied successfully with a minimum of thought. Both methods resulted in a strategy being developed as the solving evolved. A strategy is the result of combining processes and skills into some kind of order with the aim of arriving at a solution or goal. A more detailed discussion of processes, skills and strategies and their contribution to problem-solving is given by Frobisher (1994).

PROBLEMS AND MORE PROBLEMS

One of the features of process problems is that they are the source of new problems resulting either from the solving of the problem or by the asking of the questions 'What if . . .?' or 'What if not . . .?' Activity 1 is about pairs of numbers. What if it were about triples of numbers? quadruples of numbers? . . . *n*-tuples of numbers? Are there patterns and relationships to be discovered and explored between the solutions to all these problems? Here is not the place to answer these questions, but we suggest that you try to answer them for yourself in order to experience what it is like to solve process problems.

In a similar way Activity 2 gives rise to asking 'What if the grid was larger?', 'What if we counted rectangles instead of squares?', 'What if the grid was made of triangles?' Often the posing of new problems and their solutions is more interesting and challenging than the original problem. This is particularly so for students as the original process problem is often asked by the teacher, while the new problems are suggested by the students themselves. Students are able to ask further questions after exploring Activity 4 with the coins. What questions would you ask after solving this process problem?

WHAT IS AN INVESTIGATION?

Cockcroft (1982) called for the introduction of problem-solving and investigations into schools. However, there is unfortunately precious little in the report that actually explains what an investigation *is* or what it entails. What then is an investigation? Are investigations nothing more than problems with a fashionable name? If mathematics investigations are not problems, then they must have different characteristics from problems.

The three criteria used in the discussion of a problem form a useful starting point for considering the differences. If an activity does not have a specific and recognizable goal then perhaps it is an investigation, not a problem. 'Exploring square numbers' would then be classified as an investigation as it has no specific and identifiable goal. Students asked to explore square numbers need to determine the goals for themselves. The second criterion is as applicable to an investigation as it is to a problem. Unless a student responds to an investigation as a challenge it deserves to play no part in a student's education and is not worthy of its name. What about the third criterion? It follows that, as an investigation does not have a prescribed goal, then no readily known or recallable mathematics procedure is available to enable immediate progress; if there is no goal, there can be no accessible procedure. (The pupil might, however, formulate goals which then prompt the recall of relevant procedures.) It would appear that the distinction between a problem and an investigation is the existence of a clear goal specified in the statement of the activity. This distinction also explains the notion of 'open problem' which is an expression used mainly in the USA. A 'problem' could be said to be 'open' when no goal is specified, the goal being an open decision, yet to be made. A similar expression, 'open-ended problem', is also found in use in the UK, as well as other parts of the world. Here the openness would appear to refer to the problem having no end. As already shown, process problems produce many other problems to solve; this is the open-ended nature to which the name refers.

Thus an open problem is another name for an investigation, whilst an open-ended problem is a process problem which gives rise to further problems. However, in practice many teachers and textbooks use the word 'investigation' in the sense in which we have used 'process' or 'open-ended problem'. This would be acceptable if it were not so confusing for students. Do you ask your students to solve an investigation? When was the last time you asked your students to explore a problem? Students have expectations when they are presented with an activity. On the basis of the language used in an activity, whether it be presented in a written form or orally, students make assumptions about the nature of the activity. An investigation provides students with the freedom to determine the goals they wish to attain. This independence and autonomy is not possible in problems having a precise and unambiguous goal with a known and well-established method of solution.

In practice, an investigation is characterized by a spirit of dynamic engagement on the part of the investigator. There is a strong element of curiosity as the investigation moves into the unknown with many possible paths to choose and follow. Some of these paths lead to exciting and novel revelations as well as mathematics which for the child is new, others quickly lead nowhere with few conclusions reached. Children need to come to terms with the frustrations and disappointments as well as the pleasures and satisfactions when they explore new territory. It is also a major difference between an investigation and a problem. Short-term goals give satisfaction and feedback when a problem is solved. Initially students find the lack of short-term goals when working on investigations frustrating and demotivating. Unfortunately the experience of a traditional mathematics curriculum conditions students to expect such rewards to arrive quickly and frequently. Introducing investigations into the curriculum is a task not to be taken lightly even by innovative and adventurous teachers. The next chapter will consider how a gradual and planned approach to incorporating process problems and investigations into the curriculum is achievable and how teachers might adopt an investigative approach to teaching and learning.

CHAPTER 4

An investigative approach to learning

INVESTIGATIONS AND BELIEFS

Teachers of mathematics in primary and secondary schools tend to view process problems and investigations as one and the same. For this reason the term 'investigation' will be used in this chapter for both process problems and investigations as described in Chapter 3. Some teachers view mathematics as an exploration into the unknown world of symbols and are convinced that school mathematics, through its study of mathematical knowledge and skills, is a preparation for this world, not part of it. Such teachers do not recognize investigations as having a place in children's experience of mathematics until they become university students, and then only perhaps at the postgraduate level. To other teachers, mathematics comprises only knowledge and skills, and the purpose of school mathematics is to transmit these ideas to children. Watson (1983, p. 38) summarizes the views of these two groups when maintaining that 'many teachers view [school] mathematics entirely in content terms; where process skills are acquired in such cases fortuitously, since there are no specially contrived learning experiences designed to foster their development'.

There are many teachers in both these groups who are willing to discuss what an investigation might be and to consider how using an investigation with children might differ from what they are accustomed to teaching. For these teachers a step-by-step introduction of investigations into their classrooms may be the way forward. There are also a growing number of teachers who believe that investigations are an essential feature of the mathematical experience of all children and advocate that school mathematics should reflect the nature of mathematics as an exploratory activity available to all. Ernest (1991, p. 283) strongly supports this position, asserting that 'school mathematics for all should be centrally concerned with human mathematical problem posing and solving. Inquiry and investigation should occupy a central place in the mathematics curriculum.' Teachers' beliefs about the nature of mathematics and the role of investigations in the curriculum develop gradually and unconsciously, being formed over many years of mathematical experiences, planned or accidental; they cannot be changed overnight, but they can be challenged. To their surprise many teachers, after an initial reluctance and antipathy, are now finding that successful experiences 'teaching' investigations have made them reconsider their perspective of what mathematics is, and the kind of experiences they plan for children.

THE PRESENT POSITION

Despite the pressure for change which has built up over the last thirty years, different schools, both primary and secondary, are at a variety of stages along the voyage from a routine problem-solving curriculum to using an investigative approach in the classroom. The differing responses to the pressure for a change of approach to teaching mathematics are worthy of some consideration. Schools which now include real-world or practical problems as part of the mathematics curriculum view mathematics as leading to discernible outcomes. If investigations are used in these schools they are subordinate to problems or viewed merely as puzzles. Some teachers pay lip service to the role of investigations in the learning of mathematics, treating them as another topic tacked on to the existing traditional content. Other teachers see investigations as a way of involving children in thinking processes they would otherwise not experience, but mainly as a supplement to the status quo. Ernest (1991, p.289) refers to several studies having 'revealed teachers who espoused a problem solving approach but whose practices revolved around an expository, transmission model of teaching enriched by the addition of problems'. All these positions arise from a perspective of mathematics as a body of knowledge and skills to be transmitted to students, investigations being another vehicle for achieving this. The role of the teacher is crucial in any approach to teaching mathematics and a positive attitude towards, and a belief in, what is being done are essential. As Grouws and Good (1989, p.35) warn, 'it is prudent for teachers to reflect on their beliefs and to ascertain both the depth of these beliefs and the extent to which they are comfortable with the beliefs in the light of alternative conceptions'. Children quickly sense when a teacher lacks total commitment to what is being taught or to the approach used. A teacher's personal experience of, and performance with, investigations is soon apparent when children are themselves asked to investigate. To 'teach' investigations well, teachers must themselves become investigators. As Moses *et al.* (1990, p.86) stress, 'it is the teacher who establishes a classroom climate conducive to spontaneous and productive enquiry'.

IMPLEMENTING INVESTIGATIONS IN THE CURRICULUM

In recent years the focus of innovation in the curriculum, where permitted by national curricula, has been on the introduction of investigations into the curriculum. The further step towards the introduction of investigative approaches to learning and teaching mathematics in a curriculum which incorporates investigations is a leap for both child and teacher. A remodelling of the curriculum involving investigations is relatively simple when contrasted with the introduction of an investigative approach to the entire curriculum. Curriculum change, first by assimilating mathematical investigations into the content of the curriculum and then by adopting an investigative approach to learning and teaching mathematics, should not be undertaken without thought and considerable planning and preparation. Teachers wishing to take the initial step are advised to travel together with a fellow teacher of like mind as it is easy to fall by the wayside without the moral support of colleagues able to share similar experiences.

When trying investigations for the first time, teachers should be prepared for a lack of positive reaction from children. Children who have been conditioned to view 'proper' mathematics as consisting of 'sums' are likely to show opposition to investigations. 'Tell us what to do and we will do it', or 'What do we do now?', are revealing outbursts from students who, perhaps, for the very first time are asked to think for themselves. It is impossible to give the same advice to all teachers who are considering change. Some teachers may wish to dive into the deep end, whilst others desire to take things more slowly. We have found that children are responsive, initially to 'short' investigations which provide short-term rewards. This appears to match their need for repeated and recognizable success, which motivates them to attempt that which they would otherwise reject as not being like their staple diet of 'mathematics'. There are many books which provide collections of 'short' investigations, some of which have been available for many years, but are still very appropriate for the needs of those embarking on investigations for the first time (see ATM, 1969; Banwell *et al.*, 1972; Fisher and Vince, 1989; Kirkby, 1989; Kirkby and Patilla, 1987).

INTRODUCING 'SHORT' INVESTIGATIONS

A 'short' investigation which we have found appeals to children of differing ages and abilities is described in Figure 4.1. It can be presented on an investigation sheet or orally, with the first square cut by the teacher to show children how to make a start. 'Cuts, cuts and more cuts' can be used with individual children or with small groups of not more than three children. Children like to present their findings on large sheets of coloured paper and display them for others to see. They should be encouraged to talk about what they did and what they discovered with other children. There are, of course, opportunities to extend the investigation if children appear to be ready for more, by asking them to investigate what happens when they are allowed to make two straight cuts, as shown in Figure 4.2, and even three or more cuts provided children's interest remains. A second short investigation, 'They all make 9', which you may wish to try is described in Figure 4.3.

'Short' investigations introduce children to the responsibility for making and taking their own decisions. Pupils can also begin to experience how they can use processes to develop their mathematical ideas and thinking. But 'short' investigations by their very 'short' nature give children the opportunity to experience only a few of these processes.

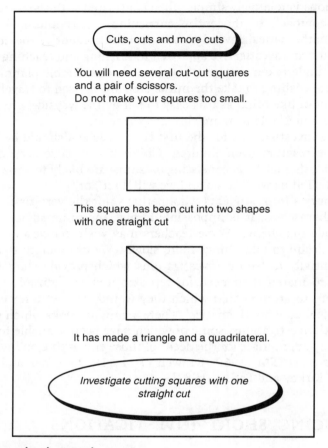

Figure 4.1   Investigating cutting squares

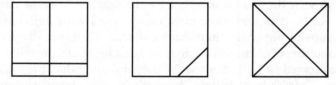

Figure 4.2   Using two cuts when cutting squares

PROCESSES IN MATHEMATICAL INVESTIGATIONS

The different kinds of processes which children may use when investigating were introduced in Chapter 3. Shuard (1986) describes a process as something we do with mathematical ideas. Frobisher (1994) classified general processes into four categories, all of which can contribute to mathematical processes which are claimed to be 'unique' to the subject (see Figure 4.4). Mathematical processes are not unique to mathematics, but play an important role in the establishment of

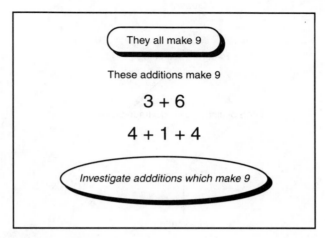

Figure 4.3 Investigating additions

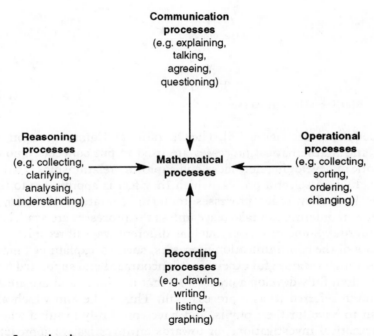

Figure 4.4 Investigative processes

new ideas and structures within mathematics. Frobisher (1994, p. 162) discusses the role of mathematical processes in investigations and suggests that the aim of a problem-centred mathematics curriculum should be to develop in children a 'knowledge of the relationships which exist between mathematical processes, as the one leads naturally into another'. He presents a sequenced list of some of the mathematical processes, to illustrate possible relationships and the order in which

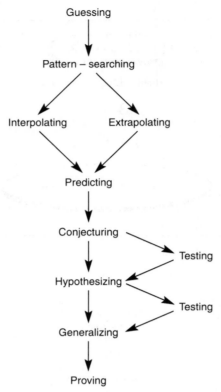

Figure 4.5   Mathematical processes

they may be related (see Figure 4.5). The Operational, Communication, Recording, Reasoning and Mathematical processes are used to put concepts, knowledge and skills to work in developing new ideas and exploring relationships. Little is known about how children select a process which they feel is appropriate to the circumstances, or how they order processes to form a strategy. However, neither a selection nor an ordering can take place unless the processes are available for selection and ordering. Some processes, such as different ways of recording, need to be taught; many of the communication processes, such as 'explaining', may be developed only through meaningful experiences, encouraged and supported by a teacher.

When children fully develop a process, so that its choice and use are automatic, then it is often referred to as a process skill. This is the aim which all teachers should wish to have for their pupils, and have constantly in mind when working with mathematical investigations, or towards an investigative approach to learning and teaching. There are, unfortunately, very many processes to be experienced and developed, and not all children will turn every process into a skill. Because there are so many processes, and so little is known about the most effective ways of learning about them, children should be introduced to working with different processes a few at a time. There is, therefore, a strong case for children exploring many different 'short' mathematical investigations before embarking on much 'longer' and more potentially rewarding investigations. Teachers will find that working with a particular mathematical investigation assists them in the recognition of which processes are likely to be encountered by children in that investigation. Over a period of time, and with exchange of experiences with colleagues, it will be possible to produce a list of the processes children use in different investigations. In this way the overall coverage of the processes should become apparent and any omissions quickly rectified.

SOME EXAMPLES OF 'LONGER' INVESTIGATIONS

We now describe in some detail two investigations which have been used successfully with children. There are dangers in doing so, as teachers, having seen how a particular investigation may develop, 'teach' the investigation to children rather than allowing children to investigate for themselves. However, we feel that teachers who have not used investigations with their children may wish to see what is possible from 'simple' beginnings.

The first 'long' investigation described in detail we call '1089 and all that!' We have used it successfully with students who are meeting investigations for the first time. You may wish to try the investigation before reading our personal account of a classroom experience with Year 6 students. The investigation can be introduced orally as a puzzle, but it is possible to use an investigation sheet like the one shown in Figure 4.6.

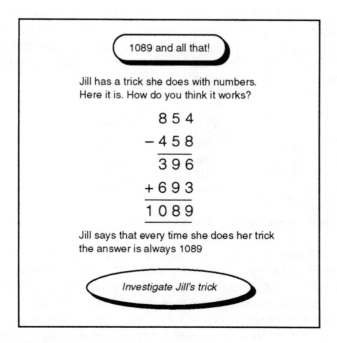

Figure 4.6   Investigating 1089: the initial situation

Table 4.1

Start numbers which work	Start numbers which do not work
871	157
936	436
331	212
614	413
472	859
593	786

Children begin by testing Jill's claim, quickly finding that it does not always work. In putting the 'trick' to the test they are producing data which are then used to investigate why it works on some occasions, but not on others. Large amounts of data are better collected if children pool what they find, rather than individuals continuing to try different three-digit starting numbers. The children are encouraged to sort their 'start' numbers as in Table 4.1. Individuals or groups are challenged to discover what is special about those 'start' numbers which work. Children not accustomed to searching for patterns or relationships do not easily notice that all the 'start' numbers which work have a hundreds digit which is larger than the units digit, and need to have their attention drawn to this fact. When this commonality is observed it is usually immediately challenged as there are numbers in the 'do not work' set to which this also applies, namely 413 and 786. Everyone is asked to look at the 'trick' with these numbers. At this stage children fall into two groups, those who find the 'end' number to be 99 and those who believe 413 and 786 have been sorted incorrectly, as they do indeed work. The difference is highlighted in calculation with 413 as the 'start' number (see Figure 4.7). Can you see the difference? The creative leap which children make in using the 'invisible' zero when reversing the 'middle' number always comes as a surprise. Everyone quickly adopts this as a rule.

$$
\begin{array}{r}
413 \\
-314 \\
\hline
99 \\
+99 \\
\hline
198
\end{array}
\qquad
\begin{array}{r}
413 \\
-314 \\
\hline
99 \\
+990 \\
\hline
1089
\end{array}
$$

Figure 4.7   Investigating 1089: starting with 413

But when is it necessary to use this new rule? Most children readily observe that if the hundreds digit is 1 more than the unit digit the rule is operable. It remains to investigate why other 'start' numbers do not work. Children who have experience of working with negative numbers produce the following calculation (see Figure 4.8) with 416 as the 'start' number. This is an interesting path to follow, but it is better left until a later stage. Decisions like this have to be made frequently in investigations, as new and different lines of enquiry keep arising.

$$
\begin{array}{r}
416 \\
-614 \\
\hline
-198 \\
+-891 \\
\hline
-1089
\end{array}
$$

Figure 4.8   Investigating 1089: starting with 416

Numbers such as 212 and 494, where the hundreds and units digits are the same, produce zero after the first reversal and subtraction. Children discuss the different types of 'start' numbers and what they produce and are left to decide on rules that must be adopted so that Jill's trick would always work. Asking students to write what the rule should be demands communicating with clarity and precision. Discussion of different suggestions provides valuable experience of the need for these concepts in mathematics. Jill's trick is amended to read 'The start number must have its hundred digit more than its unit digit', and the inclusion of the 'zero' rule applies when the two digits differ by 1.

For some children the investigation finishes here. Others wish to investigate why Jill's trick works. Once again large amounts of data need to be collected and analysed. An agreed language becomes necessary in order that everyone understands each other. In Figure 4.9, we use the language suggested by children, which all readily agree to use.

732	**START number**
−237	
495	**MIDDLE number**
+594	
1089	**END number**

Figure 4.9   Investigating 1089: start, middle and end numbers

With many calculations in front of them, the investigation focuses on the search for patterns. It is never very long before an individual or a group notices that different START numbers produce the same MIDDLE number. Try, yourself, to find three or more START numbers which produce 495 as the MIDDLE number in the calculation.

The desire to find which START numbers produce which MIDDLE numbers leads into another sort of the START numbers. The sort in Table 4.2 has already ordered the different MIDDLE numbers. This is not a natural step for most children, who tend to place the MIDDLE numbers in the order in which they appear in their many calculations. We have deliberately included only a few START numbers for each MIDDLE number, hoping that you will find many others. Child-

Table 4.2

MIDDLE numbers

99	198	297	396	495	594	693	792	891
START numbers	START numbers	START numbers	START numbers	START numbers	START numbers	START numbers	START numbers	START numbers
786	311	582	662	611	923	740	921	940
463			905	883		992	890	
						912		

dren observe many patterns in the sort, particularly that every MIDDLE number has a 9 in the middle, i.e. the ten digit of every MIDDLE number is a 9. What other patterns can you see in the MIDDLE numbers? In order to observe any patterns in each set of START numbers, it is usually necessary to have more than one in a set and occasionally to order the set. Here is a list of some START numbers which produce 396 as a MIDDLE number.

8 2 4
5 9 1
5 2 1
6 7 2
7 3 3
9 9 5

Can you see a pattern in these numbers? Can you find patterns in the other sets of START numbers? Can you explain why the patterns occur?

Figure 4.10 is a diagram one group of children devised, with considerable help, to summarize their findings. Needless to say, it is the end product of many attempts and drafts. When students are investigating, remember to take the opportunity to teach new skills and techniques which are appropriate to their needs. In this way they take on a much more meaningful role than they would if taught devoid of a relevant context.

Many mathematical problems arise as the investigation proceeds, and children have to decide whether to pursue them at the time, or to record them to follow up at a later stage. One particularly interesting question relates to how many three-digit START numbers produce any given MIDDLE number and what relationship exists between this number and the MIDDLE number. You may wish to explore this yourself.

Although the abstract concept of proof is not an easy one, teachers may find one or two children capable of appreciating the following proof when it is worked through with them (see Figure 4.11).

The '1089 and all that' investigation is just one example of many similar investigations which can be found in publications which have been produced over the last few years. Unfortunately, such publications restrict themselves to providing starting points, leaving the exploration of the possibilities to teachers and students. This can be disconcerting to those who are unfamiliar with investigations, particularly as the processes and strategies which are likely to be used when working on an investigation are seldom listed. If they were listed, it would be possible for teachers to ensure that students are given adequate coverage and ex-

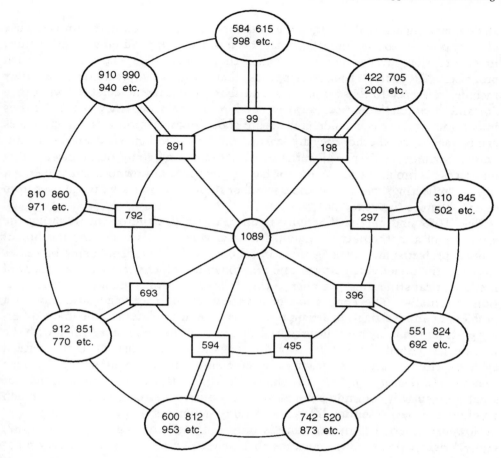

Figure 4.10   Investigating 1089: a summary

perience of those Operational, Communication, Recording, Reasoning and Mathematical processes appropriate to their stage of development.

	hundreds	tens	units
	A–1	B–1	10+C
–	C	B	A
	A – 1 – C	9	10 + C – A
+	10 + C – A	9	A – 1 – C
1	0	8	9

Figure 4.11   Investigating 1089: a 'proof'

Look back at the '1089 and all that!' investigation and try to isolate the different processes which are used. You should find you encountered the following at some stage or other: collecting data, testing rules, sorting, tabulating, ordering, searching and recognizing patterns, discussing, debating, relating, agreeing, generalizing, creating rules, defining and proving. No one investigation requires the use of every process, so it is important that those which your children use are recorded to ensure that, over a period of time, they widen their experience of such processes. Of course, this presupposes that a school's programme of study includes reference to these processes, and that all teachers understand what they are and their mathematical purpose, and recognize when they occur. This is important, as there is considerable debate about whether processes and strategies can be taught, or whether they are learned through experience in a variety of situations. Lerman (1989, p. 74) warns us that 'there is no doubt that concentrating on process is problematic. We do not have a clear idea of how to teach "generalizing" or "reflecting", nor do we know whether these processes are carried over from one problem to the next.'

The '1089 and all that!' activity is an example of children investigating the structure of a mathematical system. Much of what comes under the heading of process problems and investigations in schools is about students using processes to access the underlying patterns and relationships, which are the embodiment of mathematical structures. We refer to this as investigating a problem or investigation 'internally'. The application of a variety of processes and strategies to a problem or an investigation attempts to reveal the interrelation of the many differing parts and thus lay bare the essential elements of the structure. Armed with the knowledge gained from internally investigating a number of mathematical systems, children are more ready to observe similarities and differences between systems. As Brown (1984, p. 13) claims, 'mathematics is not only a search for what is essentially common among ostensibly different structures, but is as much an effort to reveal essential differences among structures that appear similar'.

However, internally investigating is only the beginning. As Kilpatrick (1987, p. 127) rightly points out, 'problems themselves can be the source of new problems'. This we refer to as 'externally' investigating a problem. It is the outcome of asking, 'What happens if . . .?' In the '1089 and all that!' investigation, what happens if the trick is applied to a two-digit number, or a four-, five-, six-digit number? Each of these is a system in its own right. Do the systems have anything in common? What differences are there between the systems? Do relationships exist between the different systems? These are the kinds of questions which we should be expecting children to ask after they have externally investigated a problem or an investigation. Students become problem-posers, one of the highest pinnacles of attainment of a mathematician.

Some very simple ways of moving from 'closed' tasks to 'open' tasks are given by the National Curriculum Council (1989). If you are interested in taking the idea further, then Brown and Walter (1983) provide a theoretical base with many classroom examples of how problem-posing can originate from 'closed' problems.

AN INVESTIGATIVE APPROACH TO TEACHING AND LEARNING

Investigations bring to the traditional content-oriented curriculum three distinctive features. First, they are not goal oriented; for in the statement of an

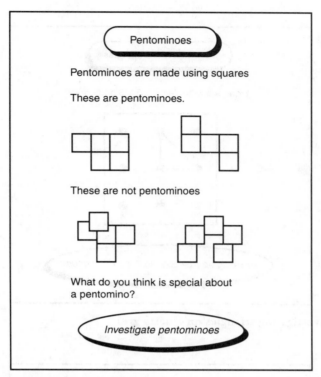

Figure 4.12   Investigating pentominoes

investigation no apparent goal is specified (see Figure 4.12). On occasions, investigations may arise out of a mathematical problem for which there is a goal, but students, in reflecting upon the problem, pose new problems. Second, investigations provide students with opportunities to decide for themselves the methods they will use, the knowledge and skills they will call upon (see Figure 4.13). Third, investigations involve students in the use of processes and strategies in their exploration (see Figure 4.14).

Investigative approaches, therefore, contain these particular characteristics together with a direct association with the 'normal' curriculum. Thus, in summary, an investigative approach to learning mathematics should:

(a) relate directly to the learning of a part of the programme of study or the set curriculum;

(b) either (i) set up mathematical learning activities or situations for which no goal is specified, or (ii) set up problems for which goals are clearly stated, but which encourage children to internally and/or externally investigate the problem;

(c) allow children to determine for themselves which knowledge and skills are needed for the learning activity; and

(d) provide children with the opportunity to use processes and strategies in their learning.

We consider how an aspect of the traditional mathematics curriculum, addition of

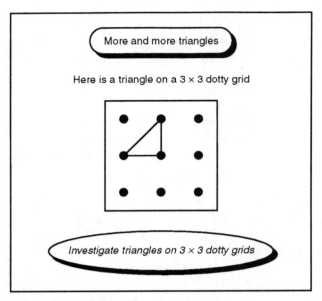

Figure 4.13   Investigating triangles on a dotty grid

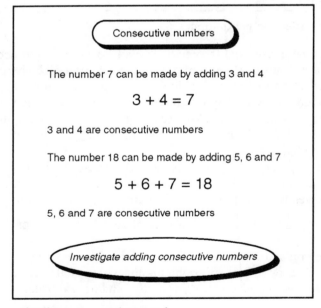

Figure 4.14   Investigating consecutive numbers

two-digit numbers, can be modified to incorporate an investigative approach. Children are presented not with additions to do, but with additions which have already been done. So instead of them being taught how to compute

47
+35
—

they are given

47
+35
82

together with similar additions and asked to explore how they work. Thus the goal is not to find answers, but to investigate methods. As part of the activity, children are expected to record 'their' method in writing and to use it on other additions which they create for themselves. They are also asked to be ready to present their work to other students. In such presentations other children are encouraged to be constructively critical, particularly asking for clarification when what is described is not understood. Different methods, including a conventional one, are compared and practised. It is a valuable learning experience to have ideas challenged by your peers, and one not easily accepted by children when experienced for the very first time.

The development of the concept of fraction involves three sub-constructs: the whole; the number of parts into which the whole is partitioned; and the equality of those parts. The concept can be developed using an investigative approach. Children are given a variety of shapes which have been partitioned (see Figure 4.15). Children sort the shapes in as many different ways as they can find, each

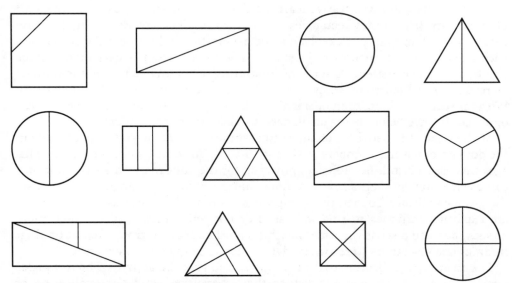

Figure 4.15 An investigative approach to the fraction concept

time describing the property or attribute they have used as the basis of the sort. Some children will notice the differing number of partitions in the shapes and use this property to sort. Their attention may need to be drawn to the equality or inequality of some of the parts in these shapes for a sort to take place using this attribute.

A third example of an investigative approach relates to the standard classroom task where students first learn the area of a rectangle. The traditional approach to teaching and learning sets numerous examples for students to practise (see Figure 4.16). This is a 'closed' problem. It is easily turned into an 'open' investigative activity for learning about areas of rectangles. Children are provided with a grid of squares. They are challenged to draw as many different rectangles as they can which have an area of 24 squares (see Figure 4.17). They are encouraged to cut out each rectangle. In this way the data, the rectangles, have a 'dynamic' form, each rectangle being separate and tangible (see Figure 4.18). The ability to move data around in order to match, compare, sort and order is of enormous benefit to mathematical thinking.

Find the area of this rectangle

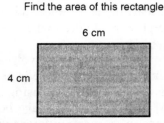

Figure 4.16   Area of a rectangle: a closed problem

It is important in investigative learning that children are not asked at the start of an activity, 'How many rectangles can you make?' The answer to this question becomes the dominant goal, replacing the exploration. Children often respond to such a question with their own question, 'How many rectangles are there?' Some children quickly experience a sense of frustration when unable to find the number of rectangles, which they believe you know, but are not prepared to tell them. Without such a goal, children operate initially at a level appropriate to them developing confidence before being challenged to make more rectangles.

At the start, it is usual for children to assume that they have to remain within the domain of natural numbers (1, 2, 3, 4, 5, . . .). To some extent the grid encourages this assumption. Be prepared, however, for the children who soon ask if they can use fractions in their attempt to make more and more rectangles which satisfy the given condition. Cooperative group work is often advantageous when students are exploring activities in this way, as they not only build up data much more quickly, but also discuss their ideas with each other. Having groups report on their findings enables others to recognize data which they may not have created.

The creation or collection of data is only the start of the activity. The rectangles form a system which has a structure that is explored with the application of processes. Matching, comparing, sorting and ordering are very relevant opera-

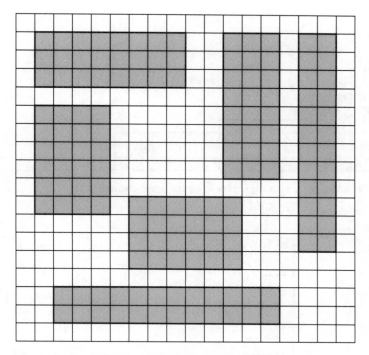

Figure 4.17   Area of a rectangle: an investigative approach

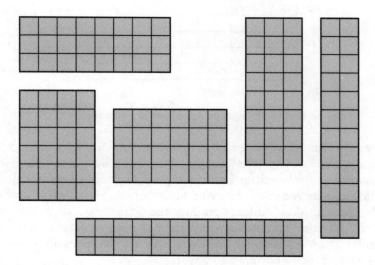

Figure 4.18   Area of a rectangle: separating the rectangles

tional processes to apply initially to a set of data. As the data are in a dynamic form, Figure 4.19 shows an outcome of sorting and ordering the rectangles which children very quickly produce. With encouragement and guidance children modify this to form a graph (see Figure 4.20).

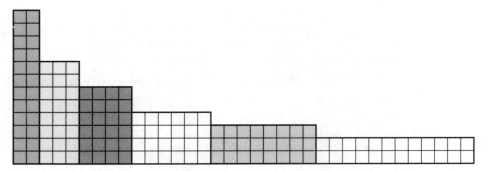

Figure 4.19   Area of a rectangle: looking for a relationship

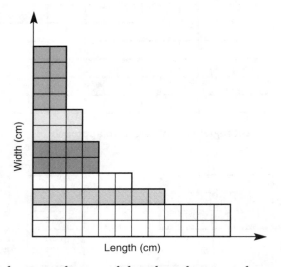

Figure 4.20   Area of a rectangle: a graph based on the rectangles

Searching for different ways of recording and displaying information is a recognizable hallmark of investigative learning. Teachers play a significant role in this particular aspect of mathematics by introducing students to the formalized ways by which data can be recorded, focusing attention on relationships which might otherwise be hidden. After students produce the 'graph' in Figure 4.20 they can be challenged to find other rectangles which have the same 'area' and to place them on the graph. They investigate the extension of the domain beyond natural numbers to halves, then quarters and beyond.

Data can, of course, be collected and recorded in a static format, usually on paper. Teachers adopting this approach should be aware that it demands considerable experience of the ways in which data are tabulated in order to bring out relationships (see Chapters 9 and 11). If students begin by drawing rectangles on the same grid without separating them, they have to recognize that the rectangles contain more informative data, the dimensions, which can be used to explore the activity much further. Some discussion may be necessary to indicate that new data are created from the original data (i.e. the rectangles), namely the length and width of each rectangle. This is an important idea in mathematics, as data in one form often give rise to 'new' data to be further explored. The lengths and widths of the rectangles are recorded in a table, at first unordered (see Table 4.3), and then ordered to assist the search for patterns and relationships (see Table 4.4).

Table 4.3

Length (cm)	Width (cm)
3	8
12	2
6	4
4	6
8	3
2	12

Table 4.4

Length (cm)	Width (cm)
2	12
3	8
4	6
6	4
8	3
12	2

Students quickly observe that they can draw a 1 × 24 rectangle, and conjecture that other dimensions may also be possible (see Table 4.5).

Table 4.5

Length (cm)	Width (cm)
1	24
2	12
3	8
4	6
5	?
6	4
7	?
8	3
9	?
10	?
11	?
12	2

The desire to close the gaps in the table leads students to look for 5 × ? rectangle. Only a few recognize this as the equation 5 × ? = 24, using the inverse operation of division to solve it. Some children return to the further drawing of rectangles to assist them in observing that the width must lie between 4 and 5. Some very sensible estimates are made on the basis of drawings. They repeat this with rectangles having lengths of 7 cm, 9 cm, 10 cm and 11 cm. Some children also suggest that it is possible to go beyond 12 cm or even beyond 24 cm, but are often uncertain about dimensions less than 1cm.

Putting the data on to a graph is the obvious next step for teachers, but children initially find this a troublesome step to make, particularly as they tend to view the 'length' domain as consisting of only whole numbers (see Figure 4.21).

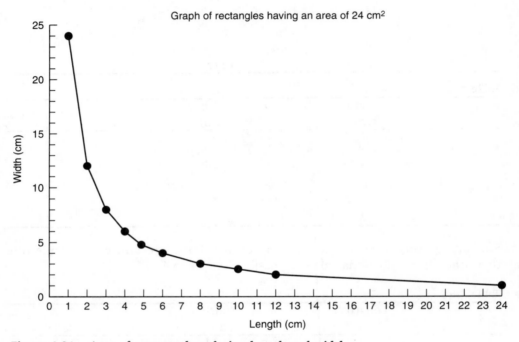

Figure 4.21   Area of a rectangle: relating length and width

As with investigations, an investigative approach gives rise to additional problems and other aspects of mathematics to explore. In the case of the 24 cm^2 rectangles, many avenues suggest themselves for further examination. What happens if the area is a different number of square centimetres? What happens if the area remains as 24 cm^2, but the shape is a triangle? The teacher and students together must decide if it is appropriate to investigate these, or alternative, questions. There is the opportunity for children to work at home on aspects which they find particularly appealing. There is little to be gained by continuing an investigative approach with children on an area of work which has lost its fascination and appeal. A condition of employing an investigative approach is that children are more than willing partners in the activities. Once this state of affairs no longer applies learning is minimal, and any further investigations are unlikely to produce mathematical dividends for the reluctant learner.

CHAPTER 5

Language and mathematics

ISSUES OF LANGUAGE

Language is the vehicle for communicating ideas and thoughts, both in talking to others and in ordering and marshalling our own thinking. It consists of 'words', but words are labels for concepts and ideas, so it is difficult to separate issues of language from issues of learning. Whenever we use a particular word, like 'triangle', 'ratio' or 'similar', we hope that it will immediately conjure up in the minds of our listeners the same meaning as the one we hold. Clearly, this does not always happen. Not only is it possible that there will be no immediate indication from our listeners that the word is in any way meaningful but also, even when it is, it might suggest a very different meaning from our own. There is enormous scope for different individuals to hold different meanings for the same word, in mathematics and in everyday speech; thus we may think we are communicating precisely when we are not. What is more, we are often unaware that the message received by someone else is not the same as the one we imparted. It should not be surprising, therefore, that teaching only by verbal exposition is often ineffective. On the other hand, precise utterances are not always necessary for unambiguous communication, because meaning can sometimes be very effectively conveyed via apparently meaningless sounds and by what has become known as body language, particularly in relation to class management and control.

In teaching mathematics, it should be clear that new words, associated with new ideas, need very careful introduction, including extensive discussion of meaning, together with opportunities for children to say, write (correctly!) and use the words of phrases in relevant contexts. It is all too easy to gloss over the introduction of new language to a class as if the desired meaning will be automatically internalized in the mind of every child. It is also unwise for a teacher, subsequently, to assume there will be no further difficulty with particular words, no matter how carefully they have been introduced, because regular use in context is essential too.

A good example of the difficulties of introducing a new concept and its associated label is that of 'real numbers', already mentioned briefly in Chapter 2. When a child has difficulty understanding what a teacher means by 'real numbers' it might be at least partly because 'number' and 'real' have already taken on particular meanings, one within mathematics and the other without. The comprehension of the specific use of 'real' and 'number' in combination requires children to make major mental readjustment and accommodation. To a novice, it could be assumed that if some numbers are 'real', then there must be others which are 'unreal', so just how do reality and unreality apply to numbers? Surely all numbers are as real or unreal as each other? The concept itself is not easy, involving as it does the gradual extension of number sets from natural numbers to integers to rational numbers and finally to real numbers (probably incorporating some relatively brief encounters with irrational numbers like $\sqrt{2}$ and π on the way), together with implications for representation on the number line. Arriving at understanding involves overcoming complications created by the particular choice of words used as a label. It almost seems as if language can interfere with, and even obscure or obstruct, the learning of mathematics simply because mathematicians have decided to adopt a particular word or combination of words, whether familiar or unfamiliar to the learner.

The issue of language in mathematics lessons is particularly critical for children whose first language is not English. There are many children in Britain for whom the language of the classroom is not the language spoken at home, and there are many more children around the world who have to cope with the even greater complexities of three or more languages. Such children need and deserve special consideration. For 'new' children, moving into a new environment and coping with a new language, it may be necessary to keep the mathematics simple at first, in order to focus on language development, and in particular on the specialist language of mathematics. For other children, however, there may be no need to simplify the mathematics. There are a number of positive strategies which the teacher can adopt, ranging from using both languages in speaking and writing, to grouping pupils with others from the same home language background but who are coping well with the language of the classroom. More detailed practical advice is given by Straker (1993).

THE VOCABULARY OF MATHEMATICS

It should be clear, from the above example of 'real numbers', that some of the language which we use in mathematics consists of words which can obviously be classified as 'Mathematical English', like 'number', and some which seem to have been taken from 'Ordinary English', like 'real', and it may not always be the mathematical words which create difficulty. The language of mathematics involves a very large vocabulary, including very many everyday words as well as the inevitable specialist terminology, sometimes referred to as the mathematical register. Considerable simplification of specialist language has taken place in recent decades, and many words considered either unnecessary, or too formidable for children, have virtually disappeared from British textbooks, for example 'multiplicand', 'subtrahend' and 'vulgar fraction'. Such words still remain in regular use, however, in many countries where mathematics is taught in English as a second language.

Even in Britain, many other technical words still remain, and teachers have to decide when and how to introduce children to them. Some words, one suspects, might be bordering on extinction through lack of regular use, like 'product' and 'quotient', and perhaps even 'numerator' and 'denominator', yet they all have their value. It is simpler and clearer to say 'quotient' than to say 'the number you finish up with as a result of dividing one number by another'! All subject disciplines use their own particular technical terms which, for those in the know, serve important functions such as providing conciseness and precision. Yet such specialist terminology can have an adverse effect on learning during the acceptance period, the time it takes to learn to live with the new word and to associate it with the appropriate meaning. Thus careful thought needs to be given to when it might be helpful, and perhaps even necessary, to use informal and familiar language like 'times' and 'share', and when it is desirable or important to introduce the more precise terminology of 'multiply' and 'divide'. Familiar language is difficult to discard, like well-used and comfortable shoes, but professional mathematicians do not talk about 'timesing', so presumably there must be a moment when it is appropriate to introduce the word 'multiply'. Loss of precision is always risky anyway, and might lead to difficulties of adjustment later. An example is the popular use of words such as 'sum' and 'histogram'. 'Sum' has many meanings, one as a verb in the instruction to add, another as a noun to describe an addition, and a third as a quantity of money. Its popular use to describe short calculations other than additions might strictly be considered incorrect, and may possibly be unhelpful to children. 'Histogram' is often misused to describe a wide variety of block graphs, and this can cause enormous difficulty later when precision is required in the study of statistics (see Chapter 11). We need to do all we can to encourage children not only to learn new mathematical words, but also then to use them with the precision intended.

Some words seem to occur within mathematics, and nowhere else, for example 'polygon' and 'isosceles', but such words are rare. This suggests they will cause problems for some children, but at least they always mean the same thing, so their origins, roots and derivations can be explained, and their meanings can theoretically be learned without confusion with usage in Ordinary English. Mostly, however, mathematicians seem to have adopted fairly common words as technical terms, which have then taken on new meaning within mathematics, for example 'difference', 'mean', 'relation', 'axes', 'origin', 'segment', 'degree', 'power', 'function', 'root', 'derivative', 'chord', and 'similar', some of which words have occurred in the previous sentence with a meaning quite different from the one in mathematics. For some of these words, the distinction between the meaning within mathematics and the meaning in everyday speech is quite subtle, the meanings being similar but not quite the same, and this creates a particular problem. A word like 'chord' presents little problem in this respect, though there might be other respects in which it does. 'Chord' has two very different uses in school lessons, one within mathematics when it means 'a straight line joining any two points on a curve', and the other within music when it describes 'a (usually) harmonious

55

collection of musical notes played simultaneously'. The two are hardly likely to cause confusion, since they are met in very different circumstances. (Another word, 'cord', which is pronounced the same, has yet other meanings, thus adding to the potential confusion.) The word 'similar', however, is used with both meanings even within mathematics, the specific distinction being that shapes which are classed as being 'similar' in the mathematical sense do not merely look somewhat alike, they have to be in exactly the same proportions. This causes great problems for many children, sometimes because we do not realize that the word has conjured up the more general meaning, and sometimes because the children just do not ever seem to grasp the specific mathematical meaning, perhaps because they are too firmly attached to the everyday meaning. There are many other words which present us with the same problem, most notably 'difference' and 'relation'.

Thus, in devoting attention to peculiar technical terminology like 'hypotenuse' and 'equilateral', we must not overlook complications associated with everyday words. There may be hundreds of very familiar simple words like 'small', 'shape' and 'size', and even 'middle', 'straight' and 'top', all used regularly within and without mathematics, and which rarely need explanation. There are other deceptively familiar words, however, like 'face', 'odd' and 'height', with which care may be needed. A cube has six 'faces' whilst we have only one, a number is 'odd' because it is not even and not because it is peculiar, and a triangle can have three different 'heights'. We have already seen in the context of real numbers how putting two words together can introduce a problem of interpretation, but the same difficulty can occur with two very familiar words. It seems, for example, that the fact that young children appear to understand 'one' and 'more' does not mean they will immediately understand 'one more' (see, for example, Thorburn and Orton, 1990). In the context of vocabulary, then, it should be clear that children need to be constantly accumulating words and their meanings, and that they need considerable help in doing so. It is perhaps surprising that vocabulary books are so rarely used in mathematics, and that glossaries in school textbooks are not common. Insufficient lesson time devoted to words and their meanings may be blamed on shortage of time, but that is no consolation to the child. There is certainly a place for specific attention to language within mathematics lessons. One possibility is for the children themselves to devise their own dictionary for use by the whole class. This will force the children to think very hard about meanings and definitions.

READING MATHEMATICS

Children need to be able to read the mathematics of learning materials, from published textbooks to handwritten workcards. This raises issues of readability. Basically, children need to be able to read and understand the text of mathematics learning materials without the particular language used acting as an obstruction or deterrent. Although measures of readability exist, they are difficult to apply to the kind of mixture of everyday language, specialist terminology and mathematical symbols which make up the pages of mathematics books and worksheets. In any case, there is often not the time to carry out a readability test. This creates a difficult problem for teachers, who are rightly urged to try to ensure that the materials they present are within the reading level of the children. How does one quickly judge the level of readability of particular materials? The factors which need to be taken into account include lengths of words, sentences and paragraphs, complexity of words, sentences and overall passages, incidence of relatively unfamiliar words and other symbols, and use of pictures, diagrams, graphs and other material which potentially can help the reader. It is easier to remember the meaning of the word 'trapezium', for example, if there is a picture of one nearby, though it is much less use if the picture is on the next page, which often happens when publishers juggle to make things fit economically. Handwritten materials tailored for particular pupils ought to be able to avoid such problems, as long as the teacher is aware of the need for the appropriate level of simplicity.

Generally, children most frequently need to be able to read mathematics when answering questions or tackling problems, and only rarely are they expected to read sections of textbooks. This seems to set mathematics apart from most other curriculum subjects, in which reading is regularly expected, and it is a pity, because we are in danger of teaching children to believe that mathematics textbooks are not usually read, except perhaps by teachers. This might be partly a reflection on what reading materials are available, but it is also our fault in that we do not educate our children to read mathematics. More should be done to insist that children do read, and to find or generate suitable textual materials for children to read. Some recent examples of booklets for children to read include the Bronto Books (Nuffield Maths 5–11, 1979) and the Today's World series by Michael Pollard *et al.* (1983).

The major use of mathematical language which most children meet is in the word problems they are expected to answer. The definition of 'word problem' has already been discussed. Word problems are familiar to pupils all around the world, and describe those questions which require the application of mathematics to achieve a solution but in which the appropriate procedure first needs to be identified from within sentences. The purpose of using word problems is usually the worthy one of trying to provide real-life settings for the application of skills or procedures. After all, it is important that children do not develop the view that mathematics is only ever met in school, and involves just the manipulation of numbers.

Many children appear to find word problems difficult. This is a cause of great concern in situations where English is not only the children's second language, but also the language used in mathematics lessons. The difficulties which children experience with word problems are largely related to language, and at least two questions follow immediately. First, is the difficulty caused because it is necessary to ensure that the appropriate skills and procedures are thoroughly understood before we introduce word problems? Second, can we make word problems more intelligible if we change the particular words and sentence structure which we use?

As regards the first, the situation is not clear-cut, because there appear to be situations, particularly those which are familiar to the children and which relate to their everyday experiences, when it can be an advantage not to divorce the mathematics from reality. This is the issue of whether the mathematics is 'embedded' in the real world, or whether it is 'disembedded' and therefore contains nothing to which children can relate, rendering it virtually meaningless. Even then there is a further complication, in that success with an embedded problem might also depend on the children being allowed to use their own method for solving this real-life problem rather than the one the teacher expects. It might be that the teacher's effort does not need to be directed to practising skills, but instead might more fruitfully be directed to a discussion of possible methods of solution.

As regards the second, we can answer much more definitely in the affirmative. A great deal of research evidence is available which should convince us that it does make a difference if you change not only the words and sentences, but sometimes the entire problem structure. Even simple mathematical situations can be tested in a wide variety of ways; for example, the order in which information is presented within a word problem can vary. Thus, short sentences which are presented in an order which matches the steps in which information is absorbed and used are ideal. Long sentences which require the child to search for information at several stages of the solution are unnecessarily difficult. It is also known that particular words can act as distractors. For example, if the word 'more' occurs within a word problem, many children will automatically add, without looking any more deeply into the situation. Correspondingly, 'less' appears to suggest subtraction. Here is an example of a question which could be presented more simply and directly.

> 'On a visit to their local sweetshop, Johann buys 8 fewer sweets than Louise. If Johann buys 18 sweets, how many sweets does Louise buy?'

A simpler form would be:

> 'Johann and Louise go to buy some sweets.
> Johann buys 18 sweets.
> Louise buys 8 more sweets than Johann.
> How many sweets does Louise buy?'

Other examples might necessitate consideration of issues raised earlier, such as lengths of words and of sentences, which clearly affect the difficulty of reading and understanding word problems. For those learning in a second language the evidence available to date suggests, not surprisingly perhaps, that children perform better when the problems are translated into their first language.

SPEAKING MATHEMATICS

Not only are children not expected to read mathematics, but the tradition of mathematics teaching is that children do not generally have much opportunity to speak mathematics, either, because most of the talking in mathematics lessons is done by the teacher – the one person in the room who needs it least, one might say. It is time this was changed. Speaking is an important element in the development of language facility, and most people find it easier to speak than to write, yet in mathematics lessons we have traditionally relied almost exclusively on the use of pencil-and-paper methods. Straker (1993) reminds us that children who experience difficulties in learning mathematics are generally offered more practice with pencil and paper, not the kind of talk and discussion with others which might be so much more valuable. Cockcroft (1982) included the recommendation that mathematics lessons should include opportunities for discussion, both between teachers and pupils and between pupils themselves. There are several reasons for encouraging children to talk (Pimm, 1987), namely:

- to communicate thoughts and ideas to others;
- to allow the teacher to gain insight into a child's thought processes; and
- to better enable the child to reflect.

Cockcroft refers to the ability of children to say what they mean and mean what they say as an important objective of teaching mathematics, just as it must be of all learning. It is certainly necessary for teachers to educate children into greater clarity and precision in the use of mathematical language, and this has been referred to earlier. Asking children to explain what they mean, either what they have said or what they have written, is at least as valuable for the child as it is for the teacher. Often, the most fruitful contexts will be concerned with solving or investigating a mathematical problem or situation, but even routine practice lessons can turn into discussions. A particularly important kind of explaining is the justifying of conclusions. If a child can persuade other children of some property or result, this is the beginnings of the important mathematical notion of proof.

A major purpose in encouraging children to talk about the mathematics they are working on is to allow them to reflect more easily and think more clearly. The act of articulating our thoughts, speaking them aloud, not only allows others to pass judgement or give a contrary opinion, but allows us to understand better what we are saying. Many is the time we solve a problem by using a friend as a sounding board, without necessarily benefiting from any response they might make. In fact, even group discussions usually include many instances of individual participants expressing thoughts which lead them to greater understanding but which are of little benefit to the other participants. Sometimes, when we ask a child to articulate their particular difficulty with the mathematics, the difficulty is resolved in the speaking.

A deliberate teaching device which encourages thinking without pencil and paper, and often encourages thinking aloud, is the use of mental arithmetic. Many years ago, mental arithmetic formed a major and almost daily part of the total diet of mathematics to which children were exposed, but it rather receded into the background as more and more teachers became concerned about the competitive, time-pressured climate which it generated in the classroom. Now teachers are quite rightly being encouraged to ensure that mental arithmetic forms a part of the curriculum, and, to tap its potential benefits to the full, not use it in any way which might destroy the confidence of the children. Mental work has much to offer. Basically, it promotes creative thinking and the use of imagery, but we have to expect not only that children might talk to themselves, but also that the outcome is often that the methods which particular children use are not the 'standard' methods or algorithms. It can therefore also be used, perhaps with an element of wider whole-class discussion, as a part of the process of developing the most appropriate written methods.

The use of question and answer is a common teaching method in mathematics lessons, but it is not a good example of what is meant by discussion. For one thing, it is under the tight control of the teacher, whose objective is often to steer pupils to the answer which is in the teacher's mind, thus reducing the task for the children to that of guessing what the teacher wants. However, at the very least it does allow the teacher to gain some insights into the children's thinking and current state of knowledge, and should help to identify some of the children who need special help. Valuable revision might also be achieved, but it will require modification if it is to contribute to the extension of children's knowledge. When a teacher uses question and answer, the responses obtained are typically brief and often lack precision, which the teacher is then tempted to clarify, but here is a wonderful opportunity to encourage the children to express their thoughts more clearly and to elaborate and extend their responses, and to involve more members of the class. Question-and-answer episodes can be developed, in this way, into something approaching discussion so that children are presented with valuable opportunities to expose their thinking, come into contact with the thoughts of others, and hence clarify their ideas.

WORKING IN GROUPS

The recommendation from Cockcroft (1982) that mathematics teaching should incorporate opportunities for discussion between pupils is usually interpreted to mean organizing teaching so as to include group activities. Different circumstances might suggest different-sized groups, but they should generally be small, perhaps involving between two and six pupils according to aims and objectives, for the obvious reason that the larger the group the less chance there is that every member will be actively involved, in discussion or in the task. It has been common for many years to find children working in pairs, but the implication seems to be that we need also, when appropriate, to set up rather bigger groups. The importance of discussion in fostering learning is underpinned by constructivist thinking (see Chapter 2). If we believe that children are not passive receivers of knowledge, but are continually involved in constructing meaning, and that meaning may need to be negotiated, this demands a forum within which such negotiation and construction can take place. Furthermore, it is possible that discussion plays an important role in the development of thinking skills and abilities. We are not surprised when young children talk to themselves, when they think and debate issues with themselves and aloud, yet we also expect that this will eventually cease to happen, when the thinking comes to be 'internalized'. Talking aloud to oneself and subsequent discussion with others are perhaps part of the same learning process.

Other benefits which have been suggested for discussion are the development of oral and language skills and the promotion of personal and social skills. None of these might at first sight appear directly beneficial in mathematical development. However, we have already seen how important it is that language is thoroughly learned as a vital part of the growth of conceptual understanding, and the development of personal and social skills is a part of the responsibility of every teacher in all subjects. Other 'affective' or attitudinal aspects of learning which can be enhanced through discussion are the improvement of confidence and the development of positive attitudes. However, it is important that the interpersonal relationships between members of a group are not strained, which has implications for the ways groups are set up (see Gibbs and Orton, 1994). Given satisfactory social relations, it is claimed that pupils learn to take turns, to listen to each other and value the opinions of others, and to receive fair criticism. Finally, it must also not be overlooked that there is evidence that more ideas are thrown up by group activity (which is not surprising really), making it more likely that problems will be successfully solved. The usual criticism of discussion is that it takes up too much time, particularly since there is such an overcrowded syllabus to be completed. The educational counter-argument is that a thorough grasp of a slightly more limited range of concepts is more likely to foster subsequent learning than being on the receiving end of an unremitting diet of exposition. The authoritative counter-argument is that the National Curriculum and Cockcroft both confirm there is an important place for discussion in mathematics lessons.

Allowing pupils to talk and discuss raises a number of problems for the teacher, not least when and how to intervene. Clearly, pupils will cease to discuss if a teacher is seen to be only too ready to interrupt with criticism or help, or even with complete answers. Yet when the teacher opts out of direct teaching, and insists that decisions and conclusions come from the group, it may seem that authority has been passed completely to the pupils. It may be thought that there are dangers and difficulties here, such as the possibility of losing 'control' of the class, and the possibility of incorrect messages about the commitment and effectiveness of the teacher being relayed to parents. There are still roles which the teacher must perform and be seen to perform, including encouraging pupils to remain on task, and not to indulge in too much social chat, and also including making the important decisions about when to call a halt and either switch to a whole-class discussion or to a short spell of direct exposition, or any other kind of activity.

It is also very helpful for both teacher and pupils if the teacher settles into each group from time to time, whilst at the same time not taking over completely. This clearly provides feedback to the teacher, but it also can be used to encourage all the various members of the group to offer ideas, it can contribute to the development of the skills of productive discussion and it allows a little judicious 'steering' if that is thought appropriate. A minor difficulty, which can be very disconcerting for new teachers, is that questions will emerge from the pupils to which the teacher does not immediately know the answer. There is no solution to this problem, and teachers have to learn to live with it.

Another important role for the teacher is in deciding on the activities on which the groups are to work. Some activities lend themselves better to discussion than others, with the manipulation of concrete objects proving particularly beneficial for children. Ball (1990) and Straker (1993) both include many examples of discussion activities.

CHAPTER 6

Representation and symbolism

INTRODUCTION

Children experience mathematics in the form of representational systems. Representations have many appearances, ranging from oral words in children's natural language, through concrete representations such as multi-base blocks, to the most abstract of representations, written mathematical symbols.

Some educators refer to any type of representation which takes the place of a mathematical idea as a symbol (Hiebert, 1988). Skemp (1971, p. 69) describes a symbol as 'a sound, or something visible, mentally connected to the idea. The idea is the meaning of the symbol.' In this chapter the term 'written symbol' is used for symbols such as '5', '+' and '=', to distinguish them from words, the symbols of a natural language.

Increasingly, over recent years, many mathematics educators have expressed concern that an important obstacle to children's learning mathematics is an untimely and overhasty introduction of written symbolism at all stages of learning, irrespective of the age and readiness of the children. As was emphasized in Chapter 5, the attaching of a name to a concept is fraught with difficulties. The further association of a symbol with a concept and its word referent provides impediments to learning mathematics. Mathematics, because of its unique written symbolism, is singularly susceptible to the detachment of symbols from the ideas which they represent. Throughout the whole of the mathematics curriculum, the names and symbols of concepts produce severe learning problems from which many children fail to recover. Dienes (1960) claims that learning difficulties with written symbols are due to the readiness of children to disengage a symbol from the concept it symbolizes, and then to proceed to manipulate the symbols, divorced from any concrete or pictorial representations. Unfortunately, teachers and textbooks encourage learners to pursue this pathway to confusion.

Written symbols are the stock in trade of mathematicians and are the essential tools for the subject's study. Symbolism permeates all areas of the subject, and without symbols mathematics would not exist as we know it and as it appears in the school curriculum. The significance of written symbols to mathematical thinking is irrefutable. Like language, both written and oral, mathematics is communicated in coded form which has to be decoded by a reader in order for it to make sense and have meaning. The code used is completely based upon conventions, with the choice of symbol often being the result of an arbitrary decision. Occasionally, it is possible to trace the origin of a mathematical symbol, where its derivation has a recognizable interpretation. Unfortunately, children have to learn the conventional meaning of each mathematical symbol separately, as seldom do relationships exist between individual symbols.

Although mathematics appears mainly in symbolic form in the classroom, aspects of mathematics are met by children in their everyday world outside school. Many mathematical words, such as 'more', 'smaller' and 'higher', are part of children's ordinary English or natural language. There is increasing evidence (Carraher, 1985) that children are extremely capable of handling mathematical ideas and calculations in everyday occurrences, which they dismally fail to do when presented with the same ideas in written symbolic form in a classroom. Thus, it appears that it is not always the mathematical ideas which cause problems. Sometimes it is the written symbolism used to represent the ideas. If this is true of aspects of the subject which children meet in familiar contexts, it does not augur well for those areas of mathematics, such as $326 \div 17$ and $79 + x = 32$, which are infrequently, if ever, met outside school.

Increasingly, over the past forty years, structured apparatus, sometimes referred to as manipulatives, has been used in the classroom to support the learning of mathematical concepts and algorithms. Manipulatives are intended to provide 'concrete' representations of the written symbols seldom experienced by children beyond the confines of school. However, only recently has it been recognized that such manipulatives are themselves representations of mathematics, and as such have embedded in them barriers to learning, comparable with the abstract symbolism they are intended to assist. Lesh *et al.* (1980, p. 50) understate the difficulties which such materials produce when they assert, 'The roles of representational systems . . . interact in psychologically interesting ways.' Perhaps, over the past four decades, an over-reliance has been placed on the importance of manipulatives in the learning of mathematics to the exclusion of other forms of representation. This point is emphasized by Lesh *et al.* when they claim that 'Psychological analyses have already shown that manipulatives are just one component in the development of representational systems, and that other modes also play a role in the acquisition and use of concepts.' (1980, p. 54). Figure 6.1, modified from Lesh *et al.* (1980), illustrates the translations among and between the differing modes of presentation. The model was first developed by Lesh in relation to rational numbers, but appears to have general applicability to the study of the learning of all mathematical concepts and their associated symbol systems.

CONCEPTS AND REPRESENTATIONS

Although mathematical concepts are abstract ideas, they begin their life for learners in the form of representations. In the very early stages of children's

63

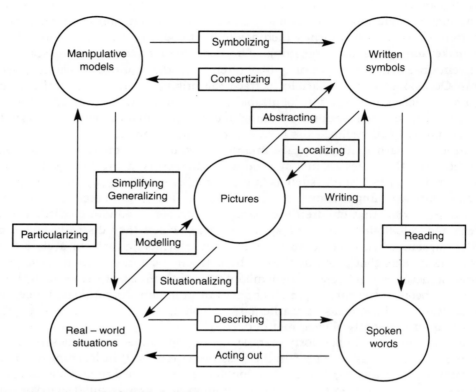

Figure 6.1   A model of symbol development
Source: modified from Lesh *et al.* (1980)

development of a concept, teachers construct contexts in which the concept is embedded. On some occasions children may have previously met the concept outside the school, and may have developed a proficiency with the idea before meeting it, perhaps more formally, in the classroom. As Carraher (1985) has shown, the level of competence in the spoken symbol system can be more advanced than that in the corresponding written symbol system taught in school. The difference in performance in the two systems, oral 'street' mathematics and written 'school' mathematics, may be due to children using different cognitive structures to develop the two systems. For example, Carraher claims that 'written representations of numeration systems are not necessary for the semantic under-standing of number to be obtained' (1985, p. 169). 'Street' mathematics con-textualizes the concepts and the context remains dominant throughout any manipulation with the ideas. The meaning of any concept is carried by the words and actions that may be activated as a consequence of the demands of the real situation. As Janvier (1989, p. 141) writes, 'numbers are processed in the opera-tions without losing their situational connotations. This equally means that the context plays an active role.' If Carraher's experience of 'street' children is gener-alizable to all learners of school mathematics then the setting of mathematics in contexts may not be helpful in the learning of written symbol systems. The strong bond between a context and the mathematics embedded within it may resist attempts to detach the mathematics and relate it to written symbols. Much work remains to be done in this important area of children's mathematics learning.

For many years teachers have placed their faith in the use of concrete representations as the source of mathematical concepts with the written symbol system as the goal. Consequently children are given many and varied experiences of concrete embodiments of mathematical concepts, the belief being that the use of concrete materials builds up mathematical imagery in a child's mind. When the appropriate imagery has been accumulated it can be used and controlled by the child without recourse to the material from which it emanated. It is as if a bridge is built between the concrete representation and the mental image of the concept. The mental image is then represented by a written symbol, enabling the child to move backward and forward between the different forms of representation. This theory claims that the concrete representation of an idea promotes both an understanding and the ready accessibility of the concept. This view is now being seriously challenged by researchers. Boulton-Lewis and Halford (1990, pp. 200–1) claim that 'It has been noted increasingly in recent literature in mathematics education that concrete representations often fail to produce the expected positive outcomes.' There is no doubt that teachers have, in the past, miscalculated the inherent problems that concrete representations create in children's learning and development of written symbols. The difficulties experienced by children in translating between concrete embodiments and written symbols are intense and are as yet hidden from those studying the interactions amongst and between different representations of mathematical ideas. Boulton-Lewis and Halford contend that 'although children can physically manipulate objects, and allocate appropriate names ("this is two", etc.), they are not recognizing the structural correspondence between concrete representations and the mathematical concept it is intended to illustrate because the load is too high' (1990, pp. 200–1).

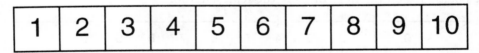

Figure 6.2   A counting strip

Other forms of representation exist which are partly within mathematics, yet are essentially pictorial in nature. The number line is an example of one such representation. Young children first meet this form of representation as a 'counting strip' (see Figure 6.2), where 'whole' numbers are placed in order, left to right (a convention to be decoded by children), in a line. The counting strip represents the ordinality of the numbers 1 to 10, but is frequently used by teachers to assist children in the answering of questions involving cardinality. Although there is nothing mathematically incorrect with the latter use, it requires children to relate two concepts of number, ordinality and cardinality, in a representational system of only one of the concepts. Thus when answering 2 + 3, using the counting strip, the numbers have to be perceived either as a position on the strip, or as a movement along the strip. The cognitive demand of such a task is difficult for children who may have only recently come to terms with the written symbolic system as a representation of the cardinality of number. It is not uncommon for children using the counting strip to find that $2 + 3 = 4$ as they count not movements of 'ones', but positions as 'ones'. The sequence of actions begins with a count of 2, 'one, two' (see Figure 6.3). This is followed by a count of 3, with the starting position of the number 2 used as the first position in the count (see Figure 6.4). Thus the number 2 is counted twice. This, from a child's point of view, is perfectly logical. The number 1 was used as the starter for the initial count of 2, and it follows that the number 2 should be the starter for the next count of 3. The counting strip is an accurate representation of the ordinality concept and provides children with a tool for manipulating numbers. However, as with all representations, it demands its own degree of understanding, both within itself and between it and the written symbol system of 'whole' numbers.

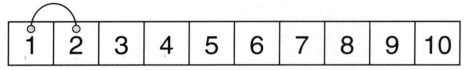

Figure 6.3 Starting addition on the counting strip

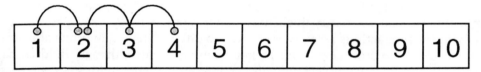

Figure 6.4 Incorrect addition on the counting strip

The counting strip should be used as a representation only of 'whole' numbers, as anomalies are quickly apparent if attempts are made to find positions for numbers such as 0, $\frac{1}{2}$, 0.642 and -4. Teachers of older children replace the counting strip with the number line, which allows every real number (see Chapter 5) to be positioned on the line. Initially the representation

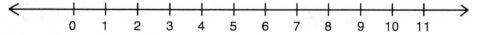

is limited to correspondence with the set of natural numbers and zero. It is assumed by many teachers that the extension to representation of rational numbers and, eventually, irrational numbers (and even integers) is natural and relatively easy for children to make. As many will testify, this is not the case. The interaction and interrelationships between the decimal symbol system and its representation on a number line, in relation to children's cognition, are not yet fully understood. A detailed description of errors and misconceptions which children develop when using the number line representation has been given by Carr and Katterns (1984). Their study involved 179 nine-year-old and 352 thirteen-year-old children using a 0 to 10 number line, with simple written symbol addition and subtraction tasks. They concluded that:

children tended to work with numerals rather than with the units or spaces between the numerals. Not only did this cause errors to be made, but it also indicated that the children did not really understand the principle on which the number line model rests. (1984, p. 34)

Representational systems carry with them their own learning difficulties because of the structures on which they are based. Unless children are conversant with the structure of a representational system there is little value in using the system to assist the understanding and operation of a different system, particularly when that system is the abstract written symbol system. Many classroom practices in mathematics lessons assume that new mathematical representations are aids to the development and understanding of concepts, yet they may in themselves be cognitively more difficult than the concepts. Mathematics educators are now searching for evidence of how children's cognitive and mathematical behaviour is determined by the interaction between representational systems. Many of these, not least the spoken or oral symbol system, are so highly developed that it is easy for teachers to minimize the difficulties for learning that such representations, and their engagement with newer and more abstract representations, demand.

Words, combinations of which form a natural language for us all, are the first representations of mathematical ideas which children meet. They are also the first symbols which children acquire and use. The decoding of oral symbols in a natural language takes place continuously as children communicate with others and participate in common experiences. Both the repeated use of words, and their immediate association with the context in which they are embedded, enable learners to develop their capacity to speak and understand others. During pre-school years children meet mathematical ideas represented by oral symbols. Only when children enter the formality of a classroom are they subjected in depth to the written symbol system of mathematics. Words in the oral system are represented by sounds, usually mentally connected to the idea. Thus children hear the word 'house' and are able to relate it to a mental or perceptual picture of a house. The sound is made up of a combination of symbols, namely letters, which often enable the sound of the word to be built up, or inferred, from the individual letter sounds. Written symbols in mathematics have sounds associated with them, but it is not possible to construct the sound if the stimulus of the visual written symbol does not provoke recall of the sound. An example of this is the symbol ':', which is used to express the ratio of two numbers, for example, 1 : 3. Thus children should learn a written symbol and its sound together, set in contexts which give meaning both to the sound and to the symbol. As Scandura (1971, p. 26) asserts, 'The arbitrary nature of the way in which symbols denote the idea makes it impossible to infer the meaning from the shape of the symbol alone.' The relationship between the two symbol systems, oral and written, is a major focus of research for mathematics educators, as children do not freely move from one representation to the other. There are many reasons for this, some of which will be discussed, as they relate to problems children have learning mathematics.

When written symbols are introduced to children for the first time they need to relate the idea to the oral representation of the idea and to the written symbol, forming a tri-linkage structural schema (see Figure 6.5). It is essential that learners of mathematics are able to recognize, read and write written symbols

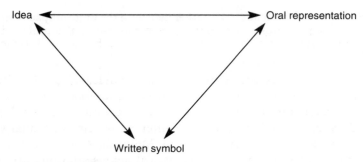

Figure 6.5   The beginning of symbolizing

appropriate to their development. This, although necessary, is not sufficient. They must be able to associate the written symbol with its corresponding concept. Children must be able to travel in both directions along all three sides of the tri-linkage schema without conscious effort. This is the challenge for teachers of mathematics, from reception classes to undergraduate level.

When children experience a concept in their everyday world its oral representation becomes part of their natural language. The related written symbol is then more readily assimilated into the linkage structure. This is particularly so when the oral representation comprises a one-syllable word and the written symbol representation is a single symbol. An example of this, which children quickly learn from birth, is the cardinality of 'one' matched to the written symbol '1'. From this point onward the relationships between idea and oral representation, oral representation and written symbol, and idea and written symbol become much more complicated, and the resulting schema more difficult to establish. The tri-linkage structure, with one slight extension (see Figure 6.6), provides a useful model for the development and evaluation of mathematical activities in relation to symbol representation. The model places emphasis on the importance of the

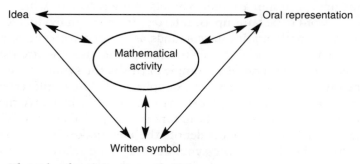

Figure 6.6   The role of activity in symbolizing

relationship between the activity and the three parts of the tri-linkage schema. If any one of the six couplings contains possible causes for errors or misconceptions the activity is likely to lead to errors and misconceptions.

Counting is one of the earliest mathematical skills children learn. In the oral symbol system sounds are matched one-to-one with 'objects' in order to determine the cardinality of the set. Children also perform activities where the written symbols, in numerical order, are placed over objects, one to each object. The two representational systems should produce the same outcome for the same set of objects. However, the two systems do not identically correspond in the way that some young children operate the oral sounds. The first six number words each have one syllable. The next word, 'seven', has two syllables. Counting for some children is associated with the correspondence between one touch of an object, or the movement of one object, and a one-syllable sound as illustrated in Figure 6.7. This possible error can be overcome by children matching the oral system directly to the written symbol, verbalizing the name of the numeral as it is placed in position, thus emphasizing the one-to-one correspondence between the two representational systems.

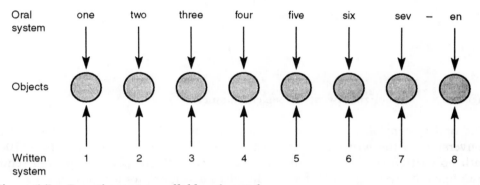

Figure 6.7 Counting: a two-syllable mismatch

Unfortunately, there are many other deficiencies of correspondence between natural language and written symbol representation. The early oral expressions representing numbers are one-word symbols. This 'oneness' of the number words is matched by the use of one written symbol up to 'nine'. The next written symbol, '10', is a combination of two previously used written symbols, '1' and '0', in that order. The number names 'eleven' and 'twelve' do not match the structure of the written symbols and do not include reference to 'ten' in any way. The lack of correspondence between oral and written representations is further complicated when the sounds of the oral symbol do not match the order of the combination of two written symbols. This is particularly noticeable with the words such as 'fourteen' and its written symbol '14'. The inversion of order is best seen when the two representations are placed beneath each other (see Figure 6.8). Most teachers of

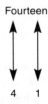

Figure 6.8 Mismatch of name and symbol

the young, and sometimes of the not so young, will have experienced children who reverse the digits of 'teen' numbers. There are further complications in the 'teen' numbers, as children correctly match the word 'ten' with the combined symbol '10', but then apply it to the part-word 'teen' in, for example fourteen. Children who apply this 'rational' approach to translating the oral numeration system to the written symbols representation for numbers proceed to write 'fourteen' as '410', which combines two errors. Children who recognize the inversion of the order, but apply incorrectly the translation of 'teen', write for 'fourteen' the written symbols '104'. A more frequently occurring error is the accurate correspondence, for example, of the word 'twenty' with the symbols '20' applied mistakenly to other numbers in the twenties (see Figure 6.9).

Twenty – five

205

Figure 6.9   Symbolizing twenty-five: a common error

The oral system of numeration also uses words which are then omitted in the conversion to the written system. Thus 'one hundred *and* four' becomes '104' with no written symbol appearing for the word 'and'. For many years some children insist on writing '1004' for 'one hundred and four', for obvious reasons.

The written representation of the numeration system does not match the oral symbols identically, leading to many possible areas of confusion. Children naively believe that mathematics is a consistent subject. Indeed they are substantially correct in this assumption, but the modes of representation contain inconsistencies which it is never possible to eradicate. A possible solution to the learning of the numeration system is a delay in the introduction of written symbols, until it is apparent that children fully understand and can apply the system orally in many and varied contexts, and that when a written symbol or combination of symbols are introduced they are directly related to their oral counterparts in meaningful situations.

Problems with lack of consistency between the correspondence of the two systems continue when children are introduced to decimals. For many years children are told that when they meet the combined symbols '27' they read them as 'twenty-seven'. This applies until written decimals such as 2.7 and 1.27 are introduced. There is some logic in reading '2.7' as 'two point seven' because of the extra symbol, the '.', separating the '2' and the '7'. It is difficult, however, to convince children that '1.27' should not be read as 'one point twenty-seven'. Indeed it is read this way in many other European countries, but with the 'point' being replaced by a 'comma', and it is read in Britain in this way when speaking of money, for example, 'one pound, twenty-seven (pence)'.

Similar inconsistencies apply in the system of rational numbers. The language of the natural number system is modified when translating into words rational numbers which appear as combinations of written symbols. For example, '²/₅' is said as 'two-fifths' and not 'two five', because of the relative positions of '2' and '5'. Children, however, are misled at the very early stages of learning about rational numbers when they are introduced to the oral words for '¹/₂' and '¹/₄'. Reading '¹/₂' as 'a half' is unquestioningly accepted by children, despite the word 'half' having no obvious interpretation. They also readily accept 'a quarter' as the expression which describes '¹/₄'. This could not go on, however, as each rational number would have its unique word form unrelated to the combined written symbols. The fractions '¹/₂', '¹/₃' and '¹/₄' have names which do not relate directly to their mathematical structure. It is also common to hear '¹/₄' spoken as 'a quarter', instead of 'one quarter', when the opportunity to emphasize the numerator is missed. A relationship does exist between the oral representation and the written symbols with 'later' fractions, such as '¹/₅' which is said as 'one-fifth'.

There are other oral symbols in mathematics which appear inconsistent with their fellows in the same system. An early example of this is the word 'eleven', mentioned earlier, for the combined written symbols '11'. For consistency with later written symbols '11' would be better described in words as 'tenty-one' (see Chapter 7).

THE NATURE OF WRITTEN SYMBOLS

Written symbols are created by men and women. They are little more than marks on paper and in themselves carry no meaning. Written symbols are arbitrary in nature and can be traced back to their origins even though changes in their shape and form may have occurred over centuries of use (Wilder, 1978). There does not appear to be a list of criteria which new written symbols should satisfy, but a brief study of the shape and form of existing ones suggests that new symbols survive if they are both simple in shape and easy to form.

Despite the capricious history of symbols and their independent and subjective beginnings, it is possible to categorize written symbols used in mathematics. Pimm (1987) suggests grouping written symbols into four main classes: logograms, pictograms, punctuation symbols and alphabetic symbols. Letters from the alphabet of different languages, lower and upper case, form the class of alphabetic symbols and are unlikely to be met by learners of mathematics until later years of schooling. Punctuation symbols are the symbols used in natural languages in printed text and are frequently used in mathematics writing. Pictograms are used in mathematics mainly in the form of geometric icons, but are not unique to mathematics, as motorists will be familiar with such symbols. Pictogram symbols in mathematics usually have a close likeness to the idea, consequently they do not present the kind of problems for learning mathematics which logograms do. One of the most humorous incidents which many teachers will have experienced occurs when children learn the symbol ∟ for a 'right angle', and then proceed to refer to a right angle when positioned in the triangle shown in Figure 6.10 as a 'left angle', even though it is at the right of the diagram. This example illustrates how children attempt to use one form of representation, in this case words which have clear and precise meanings for them, to extend their understanding of a new representation, in this case the pictogram. Such occurrences can be eliminated only if teachers make children conscious of the full meaning of the oral representation and its relationship with the written symbol.

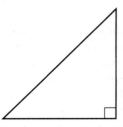

Figure 6.10   Right angle or left angle?

However, it is impossible to predict what children see as important and what irrelevant in representations. Some years ago we observed a small group of children following instructions from a textbook on the sum of the angles in a triangle. They were told to cut out a triangle and to letter the angles as shown in Figure 6.11. This they did successfully. The next instruction described and illustrated

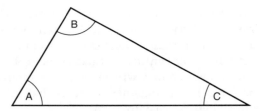

Figure 6.11   Angles in a triangle: the beginning

how to tear off each of the angles and to put them together in a special way. A diagram (see Figure 6.12) was shown. Each child did this with the three angles from their own triangle. All four children produced similar representations, shown in Figure 6.13. To the children the aspect of the representation that was important above all others was the continuous arc of the three angles as shown in the diagram in the book. Thus, when asked to write whether the angles made a straight line, all correctly wrote, 'No'. To have meaning, written symbols must be connected with meaningful referents. However, despite all the knowledge we have of how children learn mathematics, the development of activities that contain only relevant referents to which children are able to return at will escapes most teachers and textbook authors.

For a great number of years the pictorial symbol '□' represented a square. In recent years a new use has arisen, with the symbol intended to represent an unknown number in equations such as 5 + □ = 7. The 'box' symbol is much-favoured by textbook authors and has recently received a seal of approval,

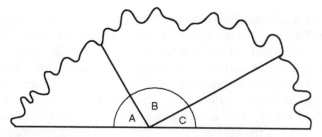

Figure 6.12   Putting the angles in a triangle on to a straight line

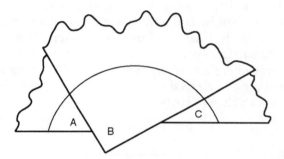

Figure 6.13   Children's perception of the angles in a triangle task

appearing in an illustrative example in the National Curriculum. Most children accept the intention that a number has to be found to 'replace' the 'box'. However, there are difficulties translating the written symbol into a word or words, for those children who do not readily recognize what they are being asked to do. When the 'box', or frame as it is sometimes referred to, appears as the first symbol, for example $\square + 3 = 9$, some children write 12 as the answer. As they are unable to give meaning to the equation, they resort to operating solely within the written symbol system, manipulating the symbols which they recognize, the '+', the '3' and the '9'. In doing so, they fail to apply the left–right direction rule of the numeration system, converting $\square + 3 = 9$ to $3 + 9 = \square$ in their desire to find an answer. If the 'box' is to be meaningful in the written symbol system, children must be able to transcribe the symbol 'into spoken mathematical language and thence into ordinary English' (MacKernan, 1982, p. 27). The most important of the classes of written symbols, logograms, is considered in the following section.

THE DEVELOPMENT OF WRITTEN SYMBOLS

The number symbols 1, 2, 3, 4, 5, 6, 7, 8, 9, 0, the operation symbols $+, -, \times, \div$ and the relation symbols $=, <, >$ are the earliest examples of logograms. They are commonly referred to as signs. Together with a straight line, —, and a dot, ., they form the sum total of most children's early experience of written symbols. Why, then, is such a small number of written symbols the source of untold problems in the learning of mathematics? The answer lies in the ways in which the finite number of symbols are combined; yet this is the power of a written symbol system, and cannot be avoided if children are to become competent in mathematics. Thus any natural number is expressed using one or more of the ten number symbols 1, 2, 3, 4, 5, 6, 7, 8, 9, 0 and a rule of combination. It is the rule of combination which creates learning difficulties. Teachers recognize the problems that the rule generates by providing children with experience of many and varied concrete and pictorial representations of how single written symbols are combined. The physical handling of a concrete representation of our 'tens' and 'ones' collective system, for example Dienes apparatus (Dienes, 1960), seldom presents children with problems. What seems to be an obstacle for children is the relationship between the concrete, or the pictorial, representation and the written symbol representation. Frequently the two representations are not seen by children to relate to one another, possibly because they are presented visually divorced from each other. Children should not be required to record in written symbols at a distance from the physical operations performed on concrete referents. At the early stages of learning about our numeration system children are encouraged to write or place written symbols close to the physical objects to which they relate (see Figure 6.14).

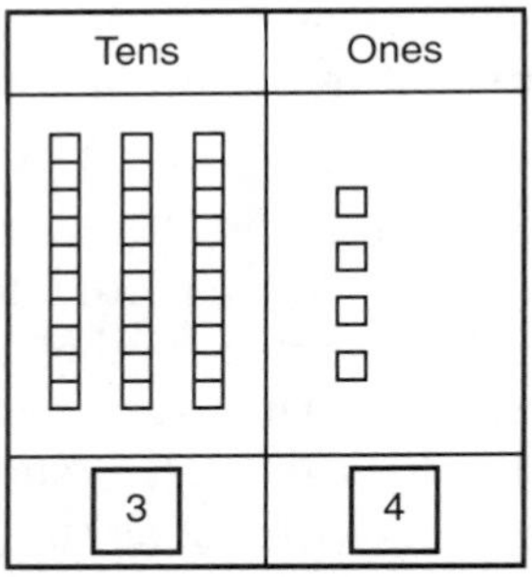

Figure 6.14   Tens and ones with structural apparatus

Repeated and abundant discussion with children is crucial in the establishment of relationships between the three representations in the above activity. It is regrettable that as children develop their mathematics, teachers sever relationships between oral symbols, concrete or pictorial representations, and written symbols before the three systems and their relationships are unshakeably entrenched as mental schemata to be operated on without conscious thought.

The order of the symbols '3' and '4' in the number 34 is an important feature of our numeration system. A change of order of the same written symbols changes the meaning of the number. Two written symbols, when used in the same order, but with a slight change in the position of one of the written symbols, produce a different meaning. Thus 34 and 3^4 have very different meanings, as have ax, a^x and a_x. It is also a convention to write some written symbols smaller when combined with other written symbols in particular ways. Indices are usually smaller than the base number; for example, a^x and a_x are more likely to be written as a^x and a_x, respectively, although it is not incorrect for children to use the same size. The orientation of a written symbol is significant for its meaning to be retained. Some young children reverse numerals, writing 3 as Ɛ. This is unacceptable, not only because of the confusion which could later ensue when written symbols are combined in complex ways, but also because communication would prove impossible. There is also the problem that the written symbol '6' is easily mistaken for a '9' when its orientation is altered. In circumstances where the orientation is determined by the way the written symbol lands, for example, on a counter or a spinner, it is customary to place a line beneath the '6' and the '9' to show which way up they are meant to be.

Written symbols which are visually similar, but have different orientations with different meanings, are agents for cognitive disorientation. Particular examples of this are the written operation symbols '+' and '×'. The orientation of a written symbol is a requisite constituent of its meaning. Order, position, size and orientation are called the surface variations of written symbols (Pimm, 1987) and, as Laborde (1990, p. 57) stresses, 'students need to acquire access to the conceptual distinctions that are marked by these surface variations'.

An ideal written symbol system would have an idea represented by a unique oral symbol, a word, which in turn would match one to one with a unique and unambiguous written symbol. A written symbol should have precision both in its form and its meaning. Regrettably, not all of the written symbol systems which represent the mathematics that children learn satisfy these criteria.

Although teachers have an essential role to play in children's constructing of and conferring meaning on written symbols, it is children who must conceptualize their own meanings and interpretations (see Chapter 2). There is ample evidence to show that children's concepts associated with some written symbols inadequately match teachers' intentions. The written operation symbol '−' is well known for its dual meaning of 'difference' and 'take away'. The latter of these is readily understood by most children, as it matches the physical actions performed on concrete representations. As a consequence of their experiences children attach to the '−' symbol the determination of how many are left after the action of removing members of a set. A well-formed schema of both representations, the concrete and the written symbol, and their relationship is constructed. Children are then presented with a further meaning for '−', where a comparison of two sets is made and the 'difference' is calculated. It is understandable when children immediately fail to model a physical representation in the form of a number 'comparison' of two sets with the same '−' symbol. A similar problem arises with the '÷' symbol where, for example, $12 \div 3 = 4$ is sometimes viewed as 'how many 4s in 12?' (known as 'grouping'), and on other occasions as 'divide 12 into 3 parts' (known as 'sharing'). To be able to use and apply a written symbol in appropriate circumstances on all occasions children should have fully operational meanings for the symbol and for the associated concepts. Children are not assisted in their development towards a fully functional schema when they are presented with alternative ways of symbolizing a concept in a variety of written ways. For example, $12 \div 3$ is also symbolized as

$$3 \underline{\smash{)}12}, \qquad 12/3, \qquad \frac{12}{3}, \qquad 3 \overline{\smash{)}12}$$

It has already been claimed that children encounter innumerable difficulties in mathematics due to the ways written symbols are combined. In many of the examples already used to illustrate the complexity of our written symbol system, combinations of symbols have been used without comment. The abundance of combinations of written symbols makes it impossible to discuss here all the problems created by such combinations. It will suffice to consider one example to show the diversity of meanings of written symbols which are possible using the written symbols '2', '4' and a single straight line. Very soon after starting school children are introduced to '4 – 2'. This is soon followed by '$\frac{4}{2}$' and '$4/2$'. Later, when children are introduced to directed numbers they meet a new meaning for '–' when used in front of a single written symbol, such as '– 2', soon to be followed by '–4 –2', where the two '–' have different meanings. Perhaps much later still, they may meet $^2\backslash_4$. The list of combinations, although finite, is so extensive that it is an achievement for children not only to remember oral representations of most of them, but also to manipulate them with some degree of success.

This is not to claim that children have an understanding of the manipulations they perform. Written symbols can be successfully manipulated by children whilst still concealing a misrepresentation or lack of understanding of the skill used. When children are unable to relate their manipulation of written symbols to other forms of representation their procedural skill becomes, as Hiebert (1988) contends, 'mechanical and inflexible'. The operations become totally detached from reality. Most computations on written symbols do not conform to the left-to-right principle that children are taught from their very first introduction to reading. The 'vertical' format for all the four operations have various orders for the reading of the appropriate symbols. Thus, children are expected not only to operate on pairs of numbers, but also to be able to remember the order of selection of the pairs and to combine the results of the selection and operations with that which follows when operating on the next pair. To us as teachers, it would be reassuring if the actions and perceptions on written symbols in such operations identically matched the physical manipulation of the concrete representations. But the complexity associated with concrete apparatus makes it impossible to arrive at a one-to-one written symbolic representation. As an exercise, you may wish to use base 10 apparatus to find the answer to 148×237, and as you do so record in written symbols the outcomes of *every* action and calculation. You will immediately recognize that the conventional 'vertical' algorithm barely resembles the concrete representation.

It is clear that if we wish children to calculate with such written symbols, then they will do so with conceptual understanding and skill only if they are able to extrapolate from simpler examples, for which the algorithm has been arrived at by association with a concrete referent. Our aim should be that, through continued application of concrete representations and written symbolic records in meaningful situations, the use of apparatus by children should be rendered unnecessary as they begin to operate solely on their mental images.

CHAPTER 7

The topic approach

INTRODUCTION

The school curriculum, particularly at the primary stage, has for many years included opportunities for children to study and learn through topic work. The justification for this practice is that, although schools usually feel compelled to organize teaching and learning within named subject compartments, such as mathematics, geography and science, knowledge cannot always be held within the boundaries of such discrete packages. In the real world, there is often no clear partitioning of knowledge. The school curriculum, indeed any national curriculum defined in separate subjects, is basically an administrative convenience, just as the division of mathematics into number, algebra, geometry and data handling is a convenience. More often than not, a thorough treatment of an educational topic, like 'Food' for example, requires study which ranges across many subjects of the standard curriculum. Thus, there are two obvious ways to ensure that coverage is comprehensive, one being to divide up the topic between subjects, and another being to study the topic as an entity, and in a cross-curricular way. It should be no surprise to find that, the younger the child, the more likely that topic work will form a regular element within the curriculum. It is at the very youngest ages that the compartmentalization of knowledge is most artificial, and when spontaneous interests of the children can, and often should, be followed up and developed immediately. In contrast, secondary school children spend most of their time studying subjects, and cross-curricular work is something of an exception, a fact which some teachers may regret.

Various descriptions are used to discuss this approach, and we have so far met two, namely 'topic' and the 'cross-curricular approach'. Other words which may be encountered are 'project', 'theme', 'centre of interest' and 'mathematics across the curriculum'. There may be differences of definition of such descriptions in the minds of some people, but it is not easy to come up with definitions which would be universally acceptable. Clearly, the cross-curricular approach implies that a variety of school subjects will be simultaneously tapped or developed in terms of knowledge, ideas, techniques and the like, and it is the kind of study which is equated with 'general theme, topic or project' in the *Mathematics Non-Statutory Guidance* (National Curriculum Council, 1989) of the National Curriculum of England and Wales. On the other hand, however, a topic might not span several subjects. Topics which are entirely within mathematics, such as 'Quadrilaterals', have an important place in both primary and secondary schools because, for example, they allow and encourage a wider view of mathematics than is often experienced in day-to-day teaching; they allow the minutiae of mathematical knowledge to be appreciated in relation to a larger whole. A broader view might be particularly appropriate at a revision or 'drawing together' stage, or it might equally well be used to provide the opportunity for investigation or research of completely new domains, perhaps with the intention of subsequently engaging in detailed study of each new area. In this chapter we wish to consider ideas solely in mathematics as well as across the entire school curriculum, and therefore we have chosen the word 'topic' in order to encompass all the possibilities.

CROSS-CURRICULAR TOPICS

The *Mathematics Non-Statutory Guidance* makes it quite clear that there is an important place for cross-curricular work both within the school curriculum and in relation to mathematics. We saw, in Chapter 1, that the first reason which is usually suggested for studying mathematics is that it is useful. Although the case for usefulness is often overstated, it is important that children do have the opportunity to experience the need to call upon mathematical ideas in solving problems in other curriculum areas. This is a justification for 'mathematics across the curriculum'. Conversely, and equally importantly, much mathematics has been developed as a by-product of the solution of real problems across the whole of human knowledge. This concept is more difficult to convey to children, but it is important to attempt to do so. It is also important that children develop an awareness of the part mathematics plays in society, and the way it impinges on their lives.

The study of topics can make a major contribution to this, because by this means children experience mathematics associated not only with other curriculum subjects, but also with cross-curricular domains of knowledge of continuing interest to themselves and relevance to society as a whole. In a cross-curricular topic, the intention is to extend knowledge and understanding in a number of school subject areas at the same time. The example of 'Food', quoted earlier, amply illustrates both how broad and extensive such a cross-curricular study might be, and how important it is to the daily lives of us all. In fact, not only are many school subjects involved in a cross-curricular study, but even the variety of elements of the mathematics curriculum which are involved is large and, in the case of food, would be likely to include some or all of the following:

- sorting and classification
- measurement of quantities

- proportions (e.g. of proteins, carbohydrates, etc.)
- comparisons, perhaps illustrated by graphs
- temperatures, °C and °F
- reading dials and scales
- recipes and ratio
- shape and size and implications for packaging and stacking
- similarity
- area and volume
- budgeting and good buys
- ordering and stock control
- distances and delivery times
- transport logistics
- distance, time and food origins.

Such a topic might occupy at least half a term, though more formal subject lessons might also be taught alongside it, so that each week consists of a mixture of topic work and more routine lessons. After all, a mixture of ways of learning is important for us all, if motivation and interest are to be maintained. Indeed, it is possible for more than one topic to be studied over the same period of time.

In an extensive school curriculum, there is a real need to economize on both time and effort whenever possible, and it is suggested in the Non-Statutory Guidance that cross-curricular topics can help. Thus, in theory, the topic of 'Maps and plans' would incorporate the study of map scales, ratio and points of the compass, which have often featured in the past within schemes of work in both mathematics and geography. Likewise, a considerable amount of work on measurement and the reading of values from scales and graphs occurs in both mathematics and science, and it would seem more economical if such work across the two subjects could be better integrated. In practice, old habits die hard, and in any case many teachers might consider it important that children meet the same curriculum elements in different contexts and at different times, in the subjects separately and also within the ambit of a more all-embracing topic. Thorough learning, after all, rarely takes place immediately, on first meeting with new ideas, and in one single context. The justification for topic work in order to economize is not as convincing as that based on the need to allow children to learn in a variety of ways, sometimes meeting new ideas on a subject basis and sometimes in a cross-curricular way, but always in the most appropriate ways for their particular stage of development. This does not rule out the possibility, however, that careful planning might lead to some economy.

Incorporating topic work into the curriculum offers a variety of benefits. Motivation and interest in the area of knowledge being studied always have the effect of enhancing the quality and quantity of what is learned, so the right topic at the right time will always lead to purposeful and productive classroom activity. Topic work also offers flexibility for the child, in that there is a less precisely defined syllabus than in most formal mathematics lessons. This therefore means that there is considerable opportunity to display individual ingenuity and creativity. The teacher will thus learn a great deal about individual children which might not come to light in more formal lessons. Mathematics is, in fact, the subject which most observers pick out as the one most frequently taught as a separate subject in formal lessons, and not integrated in any way with any other subject. Considering the interdependence of mathematics and scientific studies, this may seem surprising, yet it is a fact. Topic work is needed in order to dispel any impression which children may be inclined to develop that mathematics is only ever done in mathematics classrooms, and that it has nothing to do with the rest of life. Topic work leads to different ways of working, such as the need to research and track down information, perhaps through the use of reference books and libraries. It also allows a mixture of individual and group work, and can even be arranged so that different individuals contribute different elements to the finished whole. From the teacher's point of view, too, different ways of working may be demanded. The teacher will need to monitor progress, guide, advise and sometimes to stimulate, with direct instruction perhaps being relatively infrequent.

Topic work also brings dangers and disadvantages. It is sometimes regarded by teachers as relatively unimportant, in comparison with traditional 'subjects' like mathematics, instead of as an opportunity to teach in an integrated way with definite purposes in mind. It is sometimes used more to keep particular groups of children occupied whilst other individuals are attended to, perhaps because they have special needs in mathematics, language, or any other subject at that moment, rather than as a means of attaining carefully defined learning goals. This carries the danger of trivializing the topic, particularly if it is also undemanding. There may even be the temptation to select an undemanding topic in order to reduce the time required to oversee progress.

Sometimes, mathematics is the curriculum subject which is overlooked in topic work, in that opportunities which arise from the topic to engage in mathematical activities are not grasped. According to Burton (1994, p. 93), 'For many years, primary teachers have been embedding a lot of their work in interdisciplinary topics . . . unfortunately, mathematics was rarely incorporated . . .' This points to the need to plan topic work thoroughly, with an eye on ensuring that the balance of work across the curriculum subjects is in roughly the right proportions, that it is sufficiently demanding, and that targets and goals are first carefully identified and then subsequently attained. And from the point of view of mathematics, given the evidence of relative neglect in topic work, it is clear that the mathematical possibilities must be mapped out right from the start. Indeed, it is also essential that the intended mathematical content is neither trivial nor simple repetition, particularly of something which was included in the previous topic, and in the one before that! Children can tire of traffic surveys, and other undemanding chores which seem to crop up all too frequently; they deserve a better scheme of study which allows progress to be made in terms of the development of knowledge, skills and understanding. An example of this is that if children have carried out traffic surveys when studying 'Cars', and when the emphasis might be on simple statistics and graphing, it might not be necessary to include a traffic count when subsequently studying 'Bridges', where the emphasis might be on geometry and structures. At some much later stage, however, data from traffic surveys might be used in more advanced statistical studies which perhaps incorporate hypothesis testing.

SOME IDEAS FOR GENERAL TOPICS

There can be no end to a list of suitable topics of a general nature. Over a period of time, many teachers develop their own favourites which they use regularly, though it is important to be alert to new possibilities, particularly given that the interests of the children are vital. For completeness, we enclose a list of forty suggestions, ranging from the short and simple to the more extended and demanding:

1	Myself and friends	2	Buttons
3	Trees and flowers	4	Animals
5	My family	6	Pets
7	Dolls	8	Teddy bears
9	Up and down	10	The kitchen
11	My house	12	My school
13	Shopping	14	Cars
15	Birds	16	Television
17	My street	18	Travel
19	The supermarket	20	Food
21	Christmas	22	Holidays
23	The fairground	24	Our classroom
25	Parties	26	Clothes
27	Knitting	28	Table games
29	Team games	30	Athletics
31	Toys	32	Church
33	The Post Office	34	Communications
35	Quilting	36	Bridges
37	Shadows	38	Seesaws
39	Space	40	Time

PLANNING THE TOPIC

The selection of a topic is only the beginning. This topic may then need to be divided into major sub-themes which are appropriate for the children, and after that it is helpful to start to map out the various components on a network or topic web, to which more and more detail can gradually be added. Aims and objectives need to be taken into account from the outset, of course, even in the choice of topic, and certainly in preparing the topic web. Naturally, we should expect the children to make the first attempt at the network, with the teacher stepping in subsequently in order to stimulate and prompt further ideas. It is also necessary for the teacher to select appropriate components and targets from the National Curriculum, and to ensure that these feature on the map. The network should ideally show how the various elements are connected or related, and in particular how one element might lead naturally to another. It is at this stage that coverage of the school curriculum needs to be monitored, and steps taken to ensure that mathematics features prominently. To this end, it might be helpful to use a math-

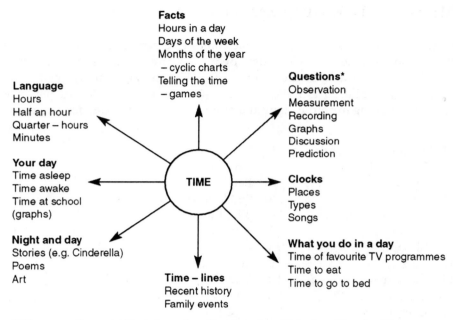

Facts
Hours in a day
Days of the week
Months of the year
– cyclic charts
Telling the time
– games

Language
Hours
Half an hour
Quarter – hours
Minutes

Your day
Time asleep
Time awake
Time at school
(graphs)

Night and day
Stories (e.g. Cinderella)
Poems
Art

TIME

Questions*
Observation
Measurement
Recording
Graphs
Discussion
Prediction

Clocks
Places
Types
Songs

What you do in a day
Time of favourite TV programmes
Time to eat
Time to go to bed

Time – lines
Recent history
Family events

[* These questions might include such as 'How long does it take to get to school?'
'How long are play-time and dinner-time?' 'How many times can we jump in a
minute?' 'How long does it take to jump ten times?' 'How long does it take all
of us to run across the playground (compare)?']

Figure 7.1   Simple network for the topic of 'Time'

ematics checklist for cross-checking with the items on the network. Such a checklist is best drawn up with reference to the particular targets and goals appropriate
to the mathematical development of the children. Figures 7.1 and 7.2 show two
different networks for the topic of 'Time', the first prepared by a teacher for a class
of young children, and the second devised by some rather older pupils. It should
be clear that both incorporate many mathematical opportunities. An example of a
mathematical checklist is as follows:

Sets	**Number**	**Measures**
Sorting	Counting	Length/distance
Matching	Order	Weight
Comparing	Addition	Capacity
Relationships	Subtraction	Volume
Patterns	Multiplication	Time
Ordering	Division	Area
	Fractions	Money
	Groupings	Relationships between sets
	More than, less than	of measures
	Relationships between	
	sets of numbers	

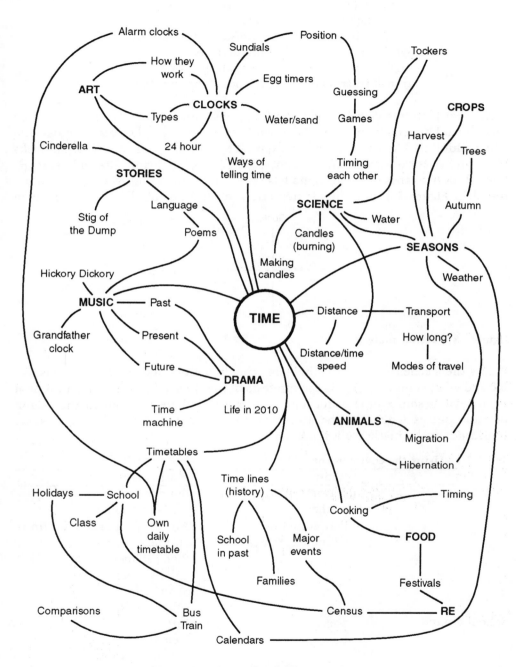

Figure 7.2   Complex network for the study of 'Time'

Shape	Data/information
Plane shapes	Collecting
Solid shapes	Tabulating
Tessellations	Pictorial representation
Symmetry	Interpreting
Angles/directions	Comparing
Transformations	Relationships between sets of data

Another planning device, which Harling (1990) informs us was introduced by the Open University in connection with its course 'Mathematics across the Curriculum', is known as 'tetrad analysis'. The four elements are **Goal(s)**, **Tasks** to be done, **Resources** required (time, equipment, etc.) and **Ground** (starting points). Each of these four elements has implications for all of the others, as illustrated in Figure 7.3, and thus systematic consideration of all the links in turn

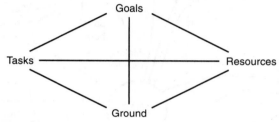

Figure 7.3   Tetrad analysis

allows the teacher to anticipate problems and difficulties and decide what is possible and what is not. In tapping the interests of the children, the local environment is an obvious source of ideas for topics. In Table 7.1 there are some further examples of sources which will be accessible to most schools, together with some indication of the mathematical content:

Table 7.1   Using the environment

Location	Item	Mathematics
Play areas	Playgrounds	Boundaries
	Pitches	Measurement
	Play activities	Arbitrary and standard units
	Climbing frames	Perimeter and area
		Plane shapes
		Solid shapes
		Structures
		Tiling patterns
School grounds	Trees	Patterning and rubbings
		Height
		Girth, growth rings
		Leaf shapes
		Leaf perimeter and area
	Shrubs/flowers	Sets and classification
		Comparison

Table 7.1 *continued*

Location	Item	Mathematics
	Buildings	Sorting
		Shape
		Tessellation
		Symmetry
		Brick bonds
		Floor patterns
	Cars	Data collection (make, colour, age)
		Registration data
		Cubic capacity
		Dimensions
	Car park	Bay size and shape
		Turning circles
		Problem-solving (e.g. how best to mark out parking areas)
Street	Signs	Shape, distance
		Speed restrictions
		Stopping distances
	Hydrants	Measurement
		Depth, number
	Houses	Types
		Odds and evens
		Place value
		Symmetry
		Plan and design
		Windows
	Lamp posts	Number
		Distance
		Height
	Markings	Straight lines
		Zigzag lines
		Parallel lines
Shop/supermarket	Costs	Range and variety
		Best buys
		Discounts
		Percentages
	Goods	Classifying
		Sorting
		Quantities
		Ingredients
Bus/railway	Timetables	Clocks
		Networks
		Distances
		Times

Table 7.1 *continued*

Location	Item	Mathematics
	Ticket office	Costs
		Savers
	Logo	Shape
		Symmetry

MATHEMATICAL TOPICS

The mathematics curriculum itself should be taught by means of a variety of approaches. There are two obvious reasons for this, namely because children need variety in order to maintain interest and motivation, and because depth and quality of learning may depend on ideas being encountered in more than one way. One might legitimately add that teachers also benefit from variety. The study of mathematics by means of the topic approach is perhaps one of the less usual ways of placing children in learning situations, but there may be occasions when it is the best approach. Two occasions have already been mentioned, one when a topic is used to integrate knowledge and ideas into a wider whole, and the other when a new area of mathematics is studied in an exploratory way, perhaps with the intention of looking in more detail at smaller elements as they arise. The topic of 'Quadrilaterals' (considered in detail below) might be used for the former reason, in order to integrate knowledge about squares, rectangles, parallelograms, etc., and the topic of 'Fibonacci numbers' illustrates the latter, because it opens up opportunities for detailed studies of so many aspects of mathematics, including the following:

- number patterns
- fractions
- formulae and equations
- the golden section
- spirals
- proof
- limits
- recurrence
- continued fractions
- number theory
- modular arithmetic
- matrices

as well as offering links with other subjects such as science (particularly the biological domain) and art. Given that Fibonacci is one of the richest areas of mathematics not formally included within the National Curriculum, one might even claim that it is the duty of mathematics teachers to find ways of including it, particularly since it opens up so many mathematical opportunities.

It should be clear that a topic which is selected in order to open up a variety of areas of mathematics has similarities with the investigative approach to learning mathematics, though perhaps on a larger scale. Topics might also offer opportunities for what has become known as 'coursework', which schools may choose to include as an element of the work that is to be assessed when pupils reach the stage of external examinations. The distinguishing feature of topics, if there is one, must be the breadth of study which it encourages. Even so, it would be possible to conceive of the topic of 'Quadrilaterals', described below, being offered as a set of investigations, or taught in an investigative way.

There are many possible mathematical topics, so many that, as with general topics, it is not possible to provide a definitive and exhaustive list. Some suggestions are, however, included here:

1	Triangles	2	Five
3	Early number systems	4	Tall and short
5	Heavy and light	6	Thick and thin
7	Circles	8	Cubes
9	Patterns in multiplication tables	10	Using squared paper
11	Dice	12	Spirals
13	Tiling and tessellations	14	Measurement
15	Angles	16	Methods of multiplying
17	Using geoboards	18	Quadrilaterals
19	Solids	20	Fibonacci numbers

'Quadrilaterals' as a topic

Quadrilaterals are plane (i.e. two-dimensional) shapes with four sides (edges), as their name tells us. The name itself need not be a stumbling-block; indeed, many children are intrigued by new and complicated words, and are keen to try them out and say them aloud. Some discussion of the derivation is appropriate, by means of splitting into 'quad' and 'lateral' . 'Quad' may then be discussed in relation to whether the children know any other words which begin in the same way (quadruple, quadruplets, quadrangle, etc.), and similar mathematical word beginnings can be compared (tri-, pent-, oct-, dec-, etc.). 'Lateral' is rather harder, but it is possible that some children will have met the word in connection with the notion of 'sideways'. Partly, but not entirely, because of the complicated language, the topic is probably appropriate for the middle years of schooling. Not all of the words present difficulties, however; for example, the name 'kite' is extremely easy to justify!

The first step is to prepare a topic web or network of related ideas which can be as comprehensive as is thought desirable (see Figure 7.4 for a possible network). The nature of mathematics is such that one idea or subtopic will inevitably lead to others, thus use of this network may involve some overlap of ideas. Some suggestions for each subtopic in the network are now included.

Sorting

A comparatively elementary activity which may only be appropriate with younger children is the sorting of plastic or cardboard quadrilaterals. One way is according to the number of square corners, another is according to the number of pairs of

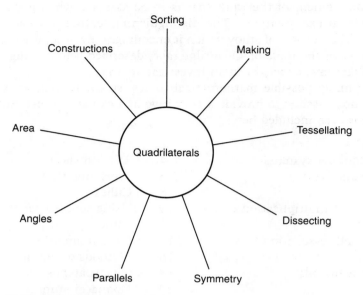

Figure 7.4 Web of quadrilateral activities

equal angles at the corners. An alternative approach is according to the number of equal sides, or according to the number of pairs of parallel sides. Such classifications will inevitably provide opportunities for profitable discussion of the differences between shapes and their respective distinguishing features. It could also eventually lead to a complete classification into square, rectangle, parallelogram, rhombus, kite, trapezium and irregular quadrilateral.

Making

All the various quadrilaterals can be cut out of paper, card, or wood, and they can be made simply by folding paper. Their skeletons can be assembled out of drinking straws and pipe cleaners (or one of the commercial equivalents), Meccano, and a number of other geometry kits. Children should experience a wide variety of different ways of making all the different shapes to help them to assimilate knowledge and ideas. The Meccano-type shapes allow transformations (rectangle to parallelogram, for example) and therefore discussion of what stays the same under such transformations (lengths? angles? area?).

Tessellating

Multiple copies of the same plastic or cardboard shapes allow investigation of how the various quadrilaterals fit together, and to the discovery of many different tessellations. The fact that particular corners fit exactly together can lead to discussion about angle sums of the particular shapes and about the sizes of the particular angles. Examples of tessellating quadrilaterals in the world outside the classroom should be sought, for example different ways of building with bricks.

Dissecting

Dissection activities often draw attention to important properties. For example, cutting a kite along its axis of symmetry reinforces angle properties and leads on to area. Cutting a rhombus along either or both diagonals does likewise. And cutting a right-angled triangle off one end of a parallelogram (and placing it at the other), or a 'half-size' right-angled triangle off each end of a trapezium, lead to methods of calculating areas. Even simpler activities like cutting along the diagonal of a parallelogram are informative in terms of properties of length, angle, area and congruence. Other simple activities include cutting a quadrilateral into two triangles, three triangles, two triangles and a quadrilateral, and so on. All such experiences enhance the children's awareness of shape and space.

Symmetry

Paper shapes allow folding and thus the investigation of the symmetry properties of all the different quadrilaterals. This leads to classification and sorting of quadrilaterals according to how many lines of symmetry they each have. The common trap which children fall into and which requires practical activity with cut-out shapes to correct is the assumption that the parallelogram has some lines of symmetry. Which shapes have 0, 1, 2, 3, 4, etc of symmetry? Which shapes have rotational symmetry?

Parallels

The study of parallel lines leads to angle properties, like the equality of alternate angles and corresponding angles. It also allows integration and reinforcement of knowledge of general angle properties and the properties of particular quadrilaterals.

Angles

The study of angles connects knowledge gained from activities associated with sorting, making, tessellating, dissecting, symmetry and parallels. Angle properties of all the shapes need to be discovered, including the sums of pairs of adjacent angles when there are parallel sides, the sum of all the angles at the vertices, the equality of particular pairs of angles, and features concerning the angles where the diagonals cross, particularly for the square, the rhombus and the kite.

Area

Methods of calculating areas are now accessible for all the various shapes, as long as the basic idea for squares and rectangles is known. The earlier dissection activities should help with the parallelogram and trapezium, and knowledge of angle properties should help with rhombus and kite. Other possible activities include the comparison of the areas of a rectangle and a parallelogram which have sides of the same two lengths.

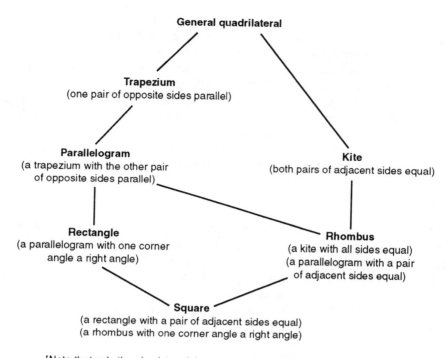

[Note that only the absolute minimum of requirements are stated. Any other properties are strictly just that – properties – and these should be recorded separately.]

Figure 7.5 Quadrilateral relationships (minimum specifications)

Constructions

Methods of drawing all the various quadrilaterals from basic minimal information help to reinforce many of the properties already discovered. Looking at it the other way round, constructing quadrilaterals allows methods of construction to be practised and learned. Examples include constructing a square of given side, a rhombus given one side and one angle, a parallelogram given two adjacent sides and one angle, and a kite given two diagonals. Discovering which constructions lead to a unique shape is part of the exercise.

Children do, of course, learn to recognize and name squares comparatively early in life, and it is squares which are therefore likely to provide the beginning for the study of all kinds of four-sided plane shapes. Other shapes and their names then follow throughout the years, with 'oblong' and 'rectangle' usually being encountered next. Authors of mathematics texts do not seem to agree on the meanings of these two words; that is, of the concepts which they signify. We shall assume that 'rectangle' is the word used to describe any four-sided shape with two pairs of parallel sides and a right angle in one corner (it does, of course, have other properties), and therefore that a square is a special kind of rectangle. In the same way, however, it can be claimed that a rhombus is a special kind of a parallelogram, that

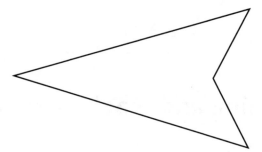

Figure 7.6   The chevron – a concave quadrilateral

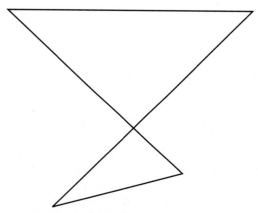

Figure 7.7   A quadrilateral with intersecting sides

a square is a special kind of rhombus, and so on. Indeed, it is thinking out these relationships and mapping the connections between the shapes which makes the topic of quadrilaterals such a good one. In order to analyse the connections, it is necessary to record the properties, in relation to sides, angles and diagonals, and in effect to come up with definitions. Studying quadrilaterals as a topic will allow this, building on the knowledge of particular individual shapes such as squares and rectangles which has accumulated over previous years. The articulation and recording of appropriate definitions for all the shapes is an important part of the topic. Children can ultimately be encouraged to produce a network diagram which shows the relations, connections and similarities between all the different quadri-laterals. One possible way of illustrating this is shown in Figure 7.5.

An unusual extension to the topic is through the idea that quadrilaterals perhaps need not necessarily be convex, and even that sides can intersect. The concave 'chevron' in Figure 7.6 alerts us to the possibility of a variety of other shapes with four sides. Figure 7.7 illustrates the idea of intersecting sides. Another possible extension is the investigation of how many different quadrilaterals of different types may be made on a geoboard of a particular size, say 4 × 4. Natu-rally, the cataloguing of quadrilaterals in the local environment needs to be included, thus providing the opportunity of moving on from a mathematical topic into a more general environmental topic. Alternatively, the topic might be used as a springboard into the broader topic of 'Polygons'.

CHAPTER 8

Learning and teaching number

PRE-SCHOOL EXPERIENCES

From birth, children are bombarded with ideas in number. The earliest experiences of number are limited to visual inputs such as the two eyes on parents' faces. Later, children begin to use the sense of touch to support and confirm that which has already been mentally accessed by the brain via visual stimuli. Parents unconsciously introduce number ideas in their everyday discourse with their offspring. Nursery rhymes, which are still taught by some mothers and fathers, contain a wealth of references to number, as well as other mathematical concepts. Prior to attending school, children are meeting number through looking at what is around them, touching and feeling objects, and listening and talking to adults and other children. None of the number concepts which develop in the first few years is the result of formal teaching, nor a consequence of seeing number represented in written symbols. Yet children are learning about number in its many forms and with its variety of meanings by listening and talking, and by experimenting and exploring in natural and meaningful situations.

It is commonly accepted that children come to attach meaning to number names through their frequent use in different contexts, although in the early stages of concept development number names are likely to be context-bound. Unfortunately, every number name has more than one meaning, the meaning becoming apparent only from the context in which it is embedded. Thus the number name 'two' is used differently when said as 'two ears' or 'the number 2 bus'. The different meanings for the same number name only very slowly become evident to children, as many teachers will confirm. This is understandable, as children not only have to distinguish between, for example, the cardinalities of 2 apples and 3 apples, and between a number 2 bus and a number 3 bus, but also to discover the ways the two uses of '2' and '3' differ and interrelate.

Number names are also used as *labels*, as for example on car number plates. Furthermore, they are used as descriptions of *quantities* such as 'four cups', and to indicate some form of *order*, as when used on pages in a book. Number words used as labels provide a useful and convenient means of differentiating and identifying objects. There is no numerical intent in their use and they could equally well be replaced by letters or other names. The shirts of a football team are frequently numbered, but players in a netball team are lettered to identify their positions on the court. In most of children's early experiences of numbers as labels, the numbers appear as written symbols, something that could, possibly, be detrimental to the more usual ways in which numbers are used, particularly in classrooms and mathematics textbooks. There is no research on how experiences of numbers as labels may influence the development of other meanings of number.

Numbers which express quantities are used in two related ways. Numbers are used to describe the cardinality or numerosity of a set of objects or events, eventually acting as a common expression for sets which may be dissimilar in nature, but for which the elements can be matched one to one. Children who have little knowledge of number names are able to lay a table, placing the necessary implements in one-to-one correspondence with the people for whom places are required. This kind of activity is well recognized as a necessary prerequisite for the understanding of cardinality. Quantities also appear when any measurement is performed. Number names are used to describe the quantity of designated units which make up a length (say 3 metres), a weight (say 97 kilograms), or money (say 37 pence). When number names are used in the context of order, three different but interrelated meanings for their use are clear. When children repeat the counting sequence of number names they do so without reference to external objects; it is a 'relatively meaningless sequence produced with some effort' (Fuson and Hall, 1983, p. 50). The sounds of the first ten number names in the counting sequence have no relationship to each other when the words are uttered. The learning of the sequence of number names occurs by repetition over many months and on many occasions. Sally, for example, came to school with the ability to say 'one, two, three, uhf, (pause), four, five, six, uhf, (pause), seven eight, nine, uhf, (pause), . . .', which was what her mother said as she climbed the stairs with Sally when going to bed. To children the counting sequence must, initially at least, appear to be a nonsensical collection of sounds. When used in an ordinal context, number names describe the relative position of an object or event in relation to the others in the ordered set. It is usual, but not essential, that the ordering begins with the number name, 'one'. Another use of number names in an order context is the most important of all number name processes, that of counting. Counting uses the number name sequence to attach to an object or event each word as it is spoken. Accurate counting demands that each object or event is paired with one and only one number name in the correct sequence.

Thus children come to school having already experienced number names used as non-numerical labels and as descriptions of cardinality and measurement, having heard them repeated in a 'meaningless' sequence which is then used to indicate relative position and, finally, having used them as a means of counting objects. Their numerical experiences of number names are inextricably linked. As far as we can tell no one helps children to recognize the differences and similarities of the different meanings of number names and the ways in which they are used. Yet, although most certainly unable to express the differences and similari-

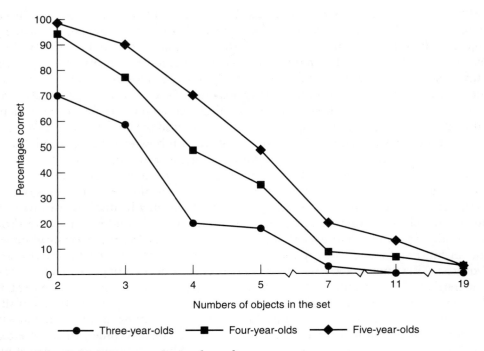

Figure 8.1   Subitizing: experimental results

ties most children prior to coming to school use the early number names confidently, consistently and correctly and apply them in a variety of contexts.

COUNTING

Without the ability to count, children's mathematical education comes to a complete halt. The DES (1979, p. 18) claims that 'the ability to count is a skill that should be acquired by almost all children before they are seven years of age'. The objective of counting is to arrive at one number name which specifies the numerosity of a set of objects, the cardinality of the set. Strangely, the process of counting, which attaches a unique number name to each object in a set, involves ordinality. There appears little doubt that as counting uses the concept of ordinality to arrive at the cardinality of a collection it has an important role in the development of both concepts. It has long been understood that children are able to recognize how many objects there are in a small collection, usually up to 5, without counting. This is referred to as 'subitizing'. Gelman and Gallistel (1978) looked at children's responses to sets of stars presented for 1 second, a time in which young children are very unlikely to be able to count. Their results are shown in Figure 8.1. The decrease in performance as the numbers increase is to be expected. As the older children will have had more number experiences than younger children, their higher success rate could also have been anticipated. It is the accuracy of the three-year-olds which is particularly revealing, as the results suggest that many have a firm grasp of the concept of cardinality, independent of counting. The results also confirm that, as the number of shapes increases, a counting strategy becomes necessary. What, then, are the sub-skills to be learned in order than children become skilful and successful counters with large numbers?

The skill of counting requires a knowledge of the sequence of number names. It begins with children choosing one of the objects in the collection to be counted as the 'starter' and naming it 'one'. Another object is then selected and named 'two'. This process continues until all the objects have been named. The last name used in the sequence represents the cardinality of the set. The choosing of objects in turn to be named requires that a record is kept of which objects have already been named and a recognition of which are available to be named.

The skill of counting operates at different levels of difficulty. One dimension of difficulty relates to the spatial positioning of the objects in the set. Counting is easier if objects are placed in a 'line' than if they are randomly positioned. When a set of objects form a 'horizontal' line, children invariably count from left to right, and have greater difficulty counting in the reverse order. Perhaps this is because reading occurs in this direction. A second dimension of difficulty is a function of the nature of the 'objects' to be counted. Children appear to find physical objects which can be moved easier to count than events, such as sounds. Teachers deliberately encourage children when counting to touch objects as they are named. However, as was pointed out in Chapter 6, 'seven' has two syllables and a child may infer the need for two touches, hence matching two objects when saying 'seven'. The touch of objects during a count hides the cardinality of the growing sub-set which has been named. When movable objects are counted, they can be physically moved as they are named, thereby making obvious the subset which has been counted and those still to be named. In this way the cardinality aspect of counting is brought to the fore. To develop the skill of counting, children should be given a variety of experiences beginning with situations which provide them with immediate success, before moving to the counting of a large number of sounds. Children can often perform the counting skill accurately, but may not have developed the concept of cardinality. This is immediately obvious when a child does not realize that whatever object is assigned the number 'one' and/or in whatever order the objects are named, the final number name used is the same.

As the skill of counting comprises a number of sub-skills, it is not surprising that there are opportunities for error. Gelman and Gallistel (1978) assessed three-, four- and five-year-old children's accuracy in counting seven relatively small sets of stars. Each child was given one minute in which to complete the count. The percentages of correct responses for each age group are displayed in Figure 8.2. It is useful to compare these results with those obtained for the same children when given only one second to respond (Figure 8.1). A substantial increase in the success rate is noticeable when children have the time in which to count even with sets which have only two, three or four objects, where subitizing would appear to be more appropriate. In recent years a substantial amount of research has focused on children's early processing of counting, and their ability to apply counting in different situations. For a more detailed discussion on children's counting the reader is referred to Fuson (1988) and Bideaud *et al.* (1992).

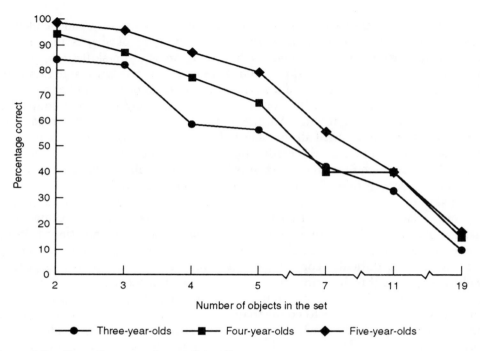

Figure 8.2   Counting: experimental results

PLACE VALUE

There is common agreement amongst teachers that 'place value' is an aspect of number which many children initially find difficult to comprehend, and later to apply in computational problems. Despite this awareness, and the readily expressed concern of teachers that children's errors in number can be traced to a lack of understanding of this concept, little progress appears to have been made in teaching place value, which would lead to a significant improvement in children's performance in this important area of the mathematics curriculum. Although children hear words such as 'twenty' and 'hundred' used in conversation, they initially perceive them as meaning 'a lot' or 'very many'. The verbal use of English number names does not immediately communicate to children the ideas of place value. Children first meet the idea of place value in the written symbol '10', the first two-digit number in the positional system. It seems, however, that the first few numbers in the second decade are viewed by children as a total of 'units', or 'ones', rather than as a combination of 'a ten and some ones'. (The word 'ones' will be used in preference to 'units', as a collection of 'ten' is also a unit.) This may be due to the construction of the oral number names which are used to represent these numbers, for example 'eleven', and 'twelve'. These words fail to carry an explicit reference to their relationship with a grouping of 'ten', nor to the earlier number names, 'one' and 'two'. In Chapter 6, the opinion was expressed that the structure of the place value system would be made more obvious to children if the words had been 'tenty-one' and 'tenty-two', or even better 'ten-one' and 'ten-two'.

This type of oral word representation which exhibits the structure of place value is common in Asian number names. In Japanese, for example, 11, 12, and 20 are spoken as 'ten-one', 'ten-two' and 'two ten(s)'. The spoken numerals in Japanese correspond exactly to their written form; plurals are tacitly understood (see Mirua and Okamoto, 1989). There are, however, implicit rules in this representation, which have to be understood to make number sense of the combined words. The number name 'ten-two' is based upon an additive structure where each part of the name is added, 'ten' + 'two', to give the number of 'ones' it represents; while the number name 'two ten(s)' uses a multiplicative rule, the 'two' and the 'ten' being multiplied, the product giving the number of 'ones'.

The written symbolic arabic numeration system, which children learn in school and which is used universally throughout the world, operates the two rules, additive and multiplicative, within all written symbols greater than 9. Thus, the number 74 comprises (7 x 10 + 4) 'ones'. Implicit in this expanded form of a number is the 'hidden' base, the collective unit of 10, upon which the system hangs. More than anything it appears that the implicit base and its powers, 10, 100, 1000, 10 000, . . . , which give position their value, are the root cause of children's learning difficulties. Although there is an obvious pattern in the positional values of the numeration system, understanding it is further complicated as place values become greater, by the continual introduction of new oral number names:

'10' is represented orally by the name 'ten'
'100' is represented orally by the name 'hundred'
'1000' is represented orally by the name 'thousand'
'1 000 000' is represented orally by the name 'million'

As numbers get larger, children learn new names of certain places in the system, whilst others combine words for their value: thus, for example; '10,000' is read as 'ten thousand'. The oral number words thus have a subsidiary base of 1000. To a degree this is also apparent in the way it is recommended that written symbols are expressed: for example 1 000 has a space, or sometimes a comma, between the 1 and the three zeros. The zero symbol, '0', is perhaps the most important, and the most difficult to understand, of all the ten written symbols, 0 to 9, which are used in the arabic numeration system. In the number '106' the zero is a place holder having 'no value'. It represents an empty place in the 'tens' position and is significant for its absence when numbers which include one or more zeros are read. Thus '106' is read as 'one hundred and six' with no reference to the number of tens. The converting of the oral representation of a number to written symbolic form is fraught with many difficulties for children. Brown (1981), as part of the Concepts in Secondary Mathematics and Science study of children's understanding of mathematics, asked twelve- to fifteen-year-olds to 'Write in figures "Four hundred thousand and seventy-three".' The facility (percentage success) at each age group was:

Age	12	13	14	15
Facility	42	51	57	57

We have tried to describe the complexities of the oral expressions used to describe large numbers and how they fail to match directly the structure of the written symbols. It is perhaps not surprising that the CSMS results are very disappointing. There is much research to be done in order to discover the exact nature of the reasons for such poor performance in this particular aspect of place value.

Most work on place value in schools concentrates on the development of children's understanding when the numbers are represented by written symbols. Ginsburg (1977) proposed a three-stage development of children's understanding of place value. Children operating at the first stage are able to write a number in written symbols correctly, but are unable to offer reasons why. During the second stage children know that there is only one way to write a number in written symbols and to recognize ways which are incorrect. For example '71' is incorrect for 'seventeen'. Children reaching the third stage are able to explain the place value notation for any given number. Although the third stage is a target that most primary teachers hope their children will reach, Ginsburg paints a depressing picture, claiming that not many move beyond Stage 2 by the age of eleven.

Resnick (1983) offers a different set of stages. Stage 1 involves the ability to partition a number into its unique expanded form: for example, 38 is 3 tens and 8 ones. Stage 2 requires the ability to produce multiple partitionings of a number, first by empirical means, perhaps by counting, and second by the direct application of exchanges, maintaining the equivalence of the number. For example, $50 + 3 = 40 + 13 = 30 + 23 = 20 + 33 = 10 + 43$, because 1 ten in turn can be exchanged for 10 ones. The final stage asks that children can apply their understanding of the previous two stages to computations, explaining how they operate. As can be seen, the models of development offered by Ginsburg and Resnick differ in major respects. There is an urgent need for a comprehensive model of children's development of place value, which models the way children operate with written symbols, and explains the errors children make and the reasons for their errors and misconceptions.

There is considerable evidence relating to children's performance on tasks which assess aspects of place value other than those already discussed. The position value of a digit was assessed by Brown (1981), APU (undated) and Kouba *et al.* (1989). The results are summarized in Table 8.1.

Table 8.1 Performance of children when asked questions relating to place value

| | | Percentage correct | | | | | |
| | | Age in years | | | | | |
Author	Question asked	$\overline{9}$	$\overline{11}$	$\overline{12}$	$\overline{13}$	$\overline{14}$	$\overline{15}$
APU (undated)	Put a ring round the number in which the 7 stands for 7 tens. 107 71 7 710		69				74
Brown (1981)	5214 The 2 stands for 2 HUNDREDS 521 400 The 2 stands for . . .			22	32	31	43
Kouba *et al.* (1989)	A. What digit is in the tens place in the number 2059? B. What digit is in the thousands place in the number 43,486?	64 45					

The questions differ in two respects. The first and last two questions require the recognition of the position of a digit given its value. The second question asks for

Table 8.2   Performances of children when asked questions relating to the size of numbers

Author	Question asked	Percentage Correct Age in Years			
		9	10	11	15
Ward (1979)	Which town has the largest population? Aberdeen 183 800 Bath 151 500 Fleetwood 28 800 Walsall 184 600 Winchester 31 000		65		
APU (indated)	Which of these numbers is the largest? 1998 2012 2004 897			91	93

Russell and Ginsburg (1984)	Which number in each pair is the larger?
	(a) 799999 (b) 522222
	        811111 288888
	(c) 833333 (d) 944444
	        177777 499999
	Success was measured by correct answers to all four questions.
		52	67		

Kouba *et al.* (1989)	Children were given four 4-digit numbers and asked for the greatest. The numbers are not specified.	70			

the value of the position given the digit. In the first and last two questions children are provided with alternatives from which to choose; these could be considered to be multiple-choice questions, and an element of guessing is possible. It would, however, seem that children are more successful in finding the position of a digit when its place value is given than when naving to state the value of a digit given its position.

Ward (1979), Russell and Ginsburg (1984), APU (undated) and Kouba *et al.* (1989) explored children's achievement on tasks involving the ordering of whole numbers. The outcomes are summarized in Table 8.2. The results suggest that the more digits the given numbers have, the more difficulty children have ordering them correctly. Many of children's experiences of numbers in real-life situations occur with relatively small numbers. Large numbers are seldom met by children outside a classroom. Thus their understanding of large numbers must be based upon representations which themselves may contain possible difficulties for learning. The use of base 10 blocks seldom ventures beyond the 'thousands', which are represented by the 'cube'. To understand numbers greater than 9999, children must transfer their understanding of the structure of numbers less than 10,000 to the larger numbers.

The limited evidence suggests that many children are unable to make this transfer successfully. Resnick (1983) conducted an interesting experiment which required nine-year-old children to count using base 10 blocks. The children were presented with sets of blocks as in Figure 8.3. The correct count would be 100,

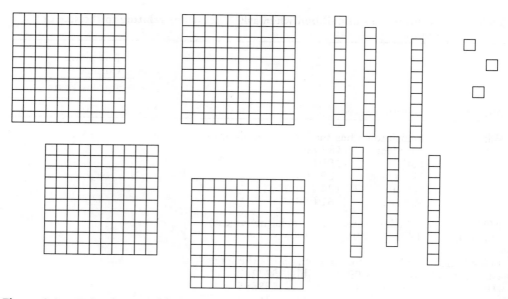

Figure 8.3   Using base 10 blocks: experimental results

200, 300, 400, 410, 420, 430, 440, 450, 460, 461, 462, 463. The mixture of different-value blocks (hundreds, tens and ones) created major problems for the children. All those interviewed were able to count only single-value blocks. The researcher interviewed a girl named Alice, asking her how much was represented by Figure 8.4. Alice responded by counting, touching the 'hundreds' as she did so,

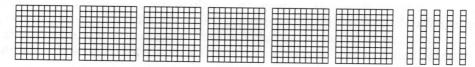

Figure 8.4   Using base 10 blocks: incorrect counting

saying '100, 200, 300, 400, 500, 600'. She continued by touching the tens, saying, '7, 8, 9, 10, 11'. On completion of her touching and counting in this way she said '611'. The 'hundred' blocks were given their appropriate value in the count, but the 'tens' took on the role of 'ones'. Alice did not appear to understand the relationship between the two different-value blocks and viewed the smaller blocks to be the unit of the system. She perceived the 'hundreds' as single entities whilst at the same time recognizing their 'hundredness', as the material clearly represents. Steffé *et al.* (1983) refer to this as treating 'hundred' as an abstract composite unit. However, the tens, although 'obviously' a collection of ten 'ones', are considered as single objects. Cobb and Wheatley (1988) cite this as perceiving 'ten' as an abstract singleton, which cognitively precedes Alice's view of the 'hundreds'. This can be explained only by the presence of both the 'hundred' and the 'ten' unit, with the 'hundred' dominating because of its position on the left of the two units.

Teaching place value using physical embodiments which represent numbers has

long been recommended (see, for example, Dienes, 1960), and adopted unquestioningly by many primary and special needs teachers. The complexities of the
place value concept are not necessarily overcome by the use of such materials.
Indeed they may bring with them added difficulties to learning which are inherent
in the nature of the material. The development of base 10 blocks and similar
materials was a consequence of an analysis of the mathematical structure of place
value by reasoning adult educators, and founded on a highly developed understanding of the concept. It is only in recent years that this has been suggested as
possibly an inadequate approach, unless supplemented by a conceptual examination of children's mathematics. This is not to suggest that base 10 blocks, abaci,
coloured counters, bundles of sticks and similar materials should never be used,
but that teachers should recognize that physical representations of number
concepts do not by themselves solve the problems for many children of learning
difficult, composite, concepts.

THE DEVELOPMENT OF EARLY IDEAS IN ADDITION

Addition has been chosen as an exemplar for all the four operations as it is not
possible to discuss in depth addition, subtraction, multiplication and division in
the space available.

Pre-school children of three and four years of age begin to form early ideas of the
meaning of addition and subtraction. These are a consequence of concrete experiences set in real situations and involve the physical actions on objects of 'adding
on' and 'taking away'. They also recognize that 'adding on' results in an increase
in the number of objects, while 'taking away' decreases the size of the set. Not
only are pre-school children developing the concepts of addition and subtraction,
but also they have the capacity to understand the outcomes of performing these
operations. Starkey and Gelman (1982) conducted an experiment involving
pennies 'hidden' in a hand to explore pre-school children's ability to calculate the
result of operating with small numbers. Table 8.3 lists the percentage successes of
sixteen children in each of the age groups three, four and five years.

Table 8.3 Performances of pre-school children in simple addition and subtraction

Age	Task 2+1	5+1	14+1	4+2	2+4	2−1	5−1	15−1	6−2	6−4
3 years (N = 16)	73	20	13	7	7	87	27	13	7	33
4 years (N = 16)	100	88	31	50	25	100	69	19	19	31
5 years (N = 16)	100	100	62	81	56	94	94	44	56	44

Source: Starkey and Gelman (1982)

Fuson (1982) and Hughes (1981, 1986), using different approaches, have
confirmed the ability of young children to perform additions and subtractions
with varying degrees of success. Most of the children in the experiments adopted

a counting strategy to add to or subtract 1 or 2 from a given number. Counting is a natural strategy for children to adopt when adding or subtracting. The strategy used varies, however, depending on how children perform their actions on the objects to be added or subtracted. Addition with movable objects, for example 3 + 1, can be accomplished in four ways. We have been unable to find any research that indicates which of the methods children favour and how the choice of method is influenced by the size and position of the numbers to be added. One or more of the methods is a child's first experience of addition, laying the foundation for its conceptual development and its eventual representation in written symbols, and it is informative to consider each method in detail, to illustrate the discrepancies which exist between that which children perceive and do, and the written symbolic record.

Method 1

The set of 3 and the set of 1 in the initial state are left unchanged in the final state when the counting takes place (see Figure 8.5). The total is counted starting with one of the sets, usually the one on the left, and continuing to the second set. This is a particularly important method to analyse as it is the only method available to children when addition is represented with objects as static pictures, which are common in worksheets and textbooks. The initial state partially matches the left side of the written symbolic representation 3 + 1 = 4, without the '+' sign. The '+' sign is often referred to by teachers as the 'putting together' action. But in this method no such physical action ever takes place; the putting together of the objects occurs only in the counting process. The '=' symbol has no match with either a physical or mental act. The '4' exists in the final state only as a mental combination of the '3' and the '1' as they retain their individual and separate identities. This approach to addition leaves children attempting to reconcile the five symbols with perceptions and actions which are not in direct one-to-one correspondence with the written symbols.

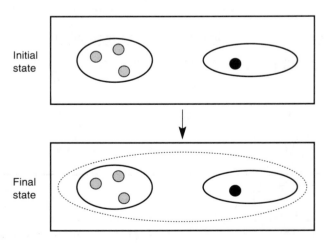

Figure 8.5   Addition: method 1

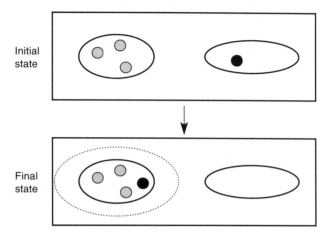

Figure 8.6   Addition: method 2(a)

Method 2(a)

The set of 3 in the initial state is transformed into a set of 4 in the final state and the set of 1 is then empty (see Figure 8.6). The '3 + 1' is not quite a representation of what occurs, as the '3' and the '1' exist in the initial state, whilst the '+' represents the action of putting the '1' with the '3'. However, the order is not '3, +, 1' but '3, 1, +' with the operator acting on the '1'. The '4' can be seen in the final state, but in the full expression, '3 + 1 = 4', there is no symbol that represents the now empty set in which the '1' was initially situated. As with method 1, the '=' sign has no match with either perceptions of quantities or with actions, physical or mental.

Method 2(b)

This is similar to method 2(a), but the objects in the set of 3 are moved to the set of 1, before the count takes place (see Figure 8.7). A child's choice of method 2(a) or 2(b) is sometimes determined on the basis of the set having either the larger or smaller number of objects.

Method 3

The 'putting together' action, when performed on the objects in both sets, suggests that the '+' sign representing the action is seen as applying to both the '3' and the '1' (see Figure 8.8). This is mathematically incorrect, but cognitively appropriate to what children do when using this method. The position of the '+' sign between the two numbers can equally imply application to both numbers, or to only the first or the second of the numbers. The method also divorces the resultant number, the '4', from the '3' and the '1'.

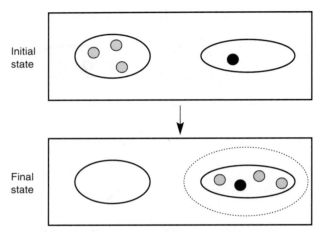

Figure 8.7   Addition: method 2(b)

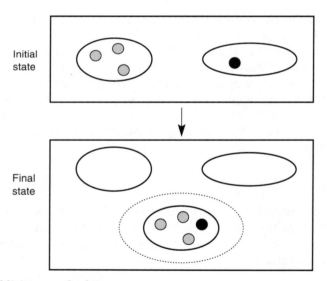

Figure 8.8   Addition: method 3

Children cannot continue for ever using objects to find the answer to an addition problem. Some of the additions which involve relatively small numbers can and should be memorized for easy recall, but teachers have difficulty stopping children counting, even when it is known that they are capable of recalling an answer from long-term memory. Additions of much larger numbers using objects are impracticable, and every child needs to have command of a way or ways of adding such numbers. Addition of single-digit numbers is commonly known as 'addition bonds' or 'addition facts'. Murray (1939) studied the relative difficulty of the 100 addition facts with 2645 Scottish children (1755 nine-year-olds and 890 eight-year-olds), and related his findings to much earlier research. The children were given the questions as a written test and, therefore, had the opportunity to work out those answers which they could not recall. Although this study was conducted many years ago, the findings, in general, have been recently confirmed in an unpublished study by Frobisher *et al.* (1993). The significant findings are summarized below:

1   The facts which have '0' as the first number, e.g. 0 + 4, are among the most difficult.
2   In general, the ties or doubles, e.g. 3 + 3, are relatively easy.
3   The size of the sum is a measure of the difficulty with, in general, larger sums being more difficult than smaller sums, and the ten most difficult non-zero combinations being 7 + 9, 8 + 9, 7 + 8, 7 + 5, 5 + 7, 9 + 7, 9 + 6, 5 + 8, 8 + 6, 7 + 6.

The results have significant implications for teaching addition facts, particularly as textbooks cannot be relied upon to give the necessary practice in the more difficult of the facts.

More recently, many researchers have explored the strategies children use to calculate answers to addition facts. Baroody (1987) provides an account of the work which has been done. Threlfall *et al.* (1995) explore whether the time taken for children either to retrieve from memory, or to calculate, can be used to determine a child's strategies. The results are inconclusive. Much of the recent work, although centring on strategies, confirms the findings from over sixty years ago, suggesting that little has changed to influence the relative difficulty of addition facts. An important implication of the recent research is how children learn relationships between facts to calculate quickly an addition fact from a known fact. For example, a child wishing to work out 6 + 5 will use knowledge of 5 + 5 and increase the answer by 1.

There is little evidence that teachers and current primary mathematics textbooks attempt to assist children in consciously developing such mathematical thinking skills. The learning of relationships between different facts should not be left to chance, although it seems that a few children do arrive at them without the assistance of teachers. Figure 8.9 illustrates ways in which some relationships can be brought to the conscious attention of children. Teachers may wish children to investigate the relationships through experiences with objects. Coloured linking cubes and dominoes provide ideal apparatus for such investigations (see Figure 8.10).

Baroody claims that, of the 100 addition facts, only the 'doubles', for example 5 + 5, need to be committed to memory. The answers to all the other facts can be quickly calculated through an understanding of the relationships which exist between the doubles and the remaining facts and the application of the commutative law, e.g. 7 + 5 = 5 + 7. Baroody refers to this process as an integration of knowledge of facts and the application of meaningful arithmetic. We could say that it teaches children to think mathematically.

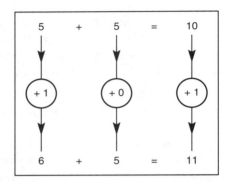 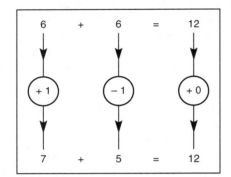

Figure 8.9   Relating additions: method 1

Cubes

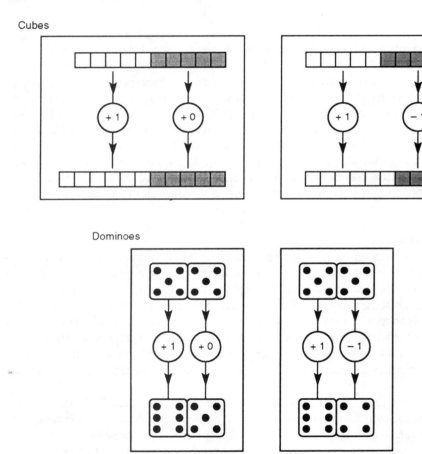

Dominoes

Figure 8.10   Relating additions: method 2

CHILDREN'S EARLY IDEAS OF FRACTIONS

We are all aware that children find fractions difficult to learn. The idea of fractional parts of a whole developed historically much later than other number

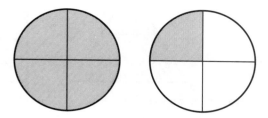

Figure 8.11   What fraction does this illustrate?

concepts, particularly their written representation in symbols, which may explain the lack of understanding that many children have of the concept and its procedures. Recent studies of achievement in the area of fractions seem to indicate that today's children are no better performers than past generations. All the evidence indicates that many children have serious misconceptions of the concept, and operate with fractions using incorrect rote procedures.

The fundamental idea behind the concept of fraction is that of a 'unit', or what is often referred to as the 'whole'. Without an understanding that a 'unit' is forever changing, and that it is a function of the situation and is not absolute, little progress can be made into the understanding of 'parts of a unit'. The idea of a unit is not new to children when they begin to learn fractions. It has already occurred in their experience, as it is the basis of the place value numeration system. However, there is a major difference, in that in place value the unit is collected together to build 'larger' units, whilst in fractions the 'unit' is decomposed into 'smaller' units. The written representation of a 'whole', using the symbol '1', is a source of confusion for some children, particularly when it is also used with another meaning in fractions greater than 1. (How can a fraction be greater than 1?) For example, Figure 8.11 represents either $1\frac{1}{4}$ or $\frac{5}{8}$ depending on the choice of the 'unit'. When children are asked to write this fraction in symbols they have no indication of the 'unit', and therefore either $1\frac{1}{4}$ or $\frac{5}{8}$ is correct.

Children are likely to meet five different meanings of the concept of fraction when learning mathematics, although the written representation will be the same for all five meanings. This many–one relationship proves difficult for teachers and children. It is common practice to introduce children first to the region idea of a fraction (this is also known as the 'part–whole' meaning). It appears that children find the 'region' model the easiest of the meanings, but whether this is because it is taught first or whether it is in absolute terms the least difficult has yet to be established. Dickson *et al.* (1984) quote Hart (1981), who found that 93 per cent of twelve- and thirteen-year-olds correctly shaded two-thirds of a rectangle already partitioned into three equal parts.

The 'region' model is directly associated with the concept of area and demands spatial perception skills, both of which may prove barriers to learning fractions for some children. For example, all the shapes in Figure 8.12 have three-quarters shaded. Each shape is considered to be a 'whole' or 'unit', which is partitioned into four parts equal in area, of which three parts are shaded. Immediately the sub-divisions are counted, they take on the role of 'units', which is a source of

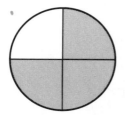 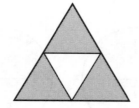 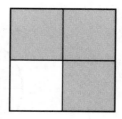

Figure 8.12   Illustrating three-quarters

confusion for some children. Although the shaded part of each of the three shapes is referred to as 'three-quarters', none of the areas in one shape is equal to any of the other shape areas. Thus the fractional parts, whilst not being 'equal', are written symbolically in an identical way.

When children are first introduced to the region model of a fraction, it is essential that reference is made to the 'whole' or 'unit'. Teachers are advised to say 'Three-quarters of this whole is shaded', rather than 'Three-quarters is shaded'. Initially, some children consider the fractional part bound to the shape of which it is a part. It is a stage in development to move from particularizing with one shape to generalizing with many different shapes.

There are three different types of activity which children should experience with the 'region' model. The first activity has already been described, and involves children being given a shape already subdivided into equal parts. They are asked to shade a fraction of the whole. As Hart (1981) discovered, the majority of twelve-year-olds are capable of successfully achieving this task (see Figure 8.13). However, we must not be misled into believing that this is an indicator of their understanding of a fraction, as all that a child has to do is to shade the number of parts equal to the first number in the expression 'two-thirds'. The task becomes merely a trivial counting activity for insightful children. Teachers may wish to try the two tasks described in Figure 8.14 with their children. The first task has more subdivisions than the partition specified in the 'two-thirds'. The second task has non-congruent subunits.

The second type of activity provides children with a whole shape, but without any subdivisions. Children have now to recognize the number of subdivisions into which they must partition the 'unit' before shading the required number of parts. This demands that they recognize and apply correctly the two aspects of a fraction (see Figure 8.15).

A third type of activity gives children a shape already partitioned and shaded. They are asked to name the fraction shaded. This is the reverse of the first type of activity. The APU (1980b) found that 70 per cent of eleven-year-olds and 80 per cent of fifteen-year-olds were correctly able to name the fraction of a circle which was shaded. Dickson *et al.* (1984) again refer to Hart's work, which found that 79 per cent of twelve- and thirteen-year-old children named the fraction shaded in Figure 8.16 as ³/₈. This result confirms the APU findings. However, Dickson *et al.* make the point that 'the areas were not only equal in size but also congruent in shape'. No evidence is available on the effect of naming fractions when the shape has non-congruent but equal-area subdivisions (see Figure 8.17). The more general 'part–whole' model encompasses other mathematical concepts as well as

'Shade two-thirds of this shape'

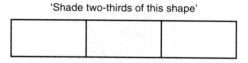

Figure 8.13   Shading two-thirds: a simple task

'Shade two-thirds of this shape'   'Shade two-thirds of this shape'

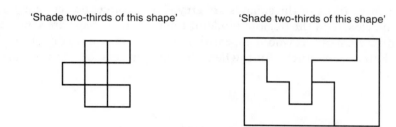

Figure 8.14   Shading two-thirds: a more difficult task

'Shade two-thirds of this shape'   'Shade two-thirds of this shape'

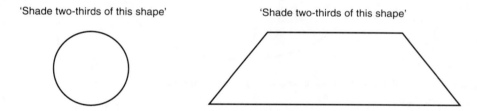

Figure 8.15   Shading two-thirds: a more open task

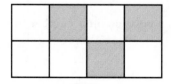

Figure 8.16   Naming a fraction: a simple task

area, which can be used as vehicles for representing fractions. These include length, weight, capacity, time, money and volume. In order to develop the 'part–whole' concept of fraction, children should be given experiences of fractions at work in all of the measures. For example, children using Plasticine and a balance should be challenged to construct a ball of Plasticine which is one-third of a given weight.

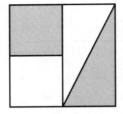

Figure 8.17   Naming a fraction: a more difficult task

A second representation of a fraction relates to the use of *discrete objects*. In Figure 8.18 three-fifths of the squares are grey. The 'discrete objects' model can be seen to relate closely to the 'region' model if the parts of a shape are separated and viewed as discrete parts. It would appear that children may have difficulty perceiving the 'whole' as a single unit when discrete objects or pictures are used.

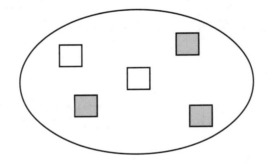

Figure 8.18   A fraction task based on discrete objects

Figure 8.19   A fraction as a position on a number line

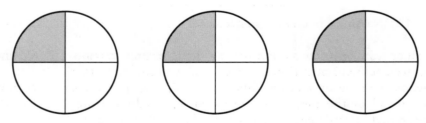

Figure 8.20   The quotient model of fraction

However, Dickson *et al.* (1984, pp. 279–80) remark that 'Novillis (1976), in her study of the hierarchical development of various aspects of fraction among American ten- to twelve-year-olds, found that the two aspects ("part–whole" and "discrete objects") were roughly of the same difficulty.'

These results contrast with the findings of Payne, again quoted by Dickson *et al.*, who found the 'sets' model to be very much more difficult than the area approach. In a practical task, the APU (1980a) found that 64 per cent of eleven-year-olds, when given three yellow tiles and one red tile, were able to state that one-quarter of the tiles were red. It would appear that more research is necessary to determine which of the models is hierarchically the more difficult and thus should be taught at a later stage in children's development of the concept of fraction.

A fraction as a *position on a number line* is a third meaning which children should experience in their development of the fraction concept (see Figure 8.19). This model treats a fraction as an abstract real number, with fractions extending children's ideas of whole numbers, including zero. The number line representation of a fraction has little association with the idea of a fraction being part of a whole, and although it is an appealing approach it would appear from evidence that fractions on a number line should not be introduced to children until well into the secondary stage (Dickson *et al.*, 1984).

Children experience a fourth meaning of a fraction when they divide, for example, 3 by 4, i.e. $3 \div 4$, for which the answer is not a whole number. This is often referred to as the *'quotient'* model. Dickson *et al.* illustrate how difficult an idea this is, when diagrams are used to represent the process of dividing 3 by 4 (see Figure 8.20). The 3 has to be recognized as three units, or three wholes, which are partitioned into four parts. As it is impossible to partition the three wholes into four wholes each of the wholes is so split. Mathematically, this is equivalent to $3(1 \div 4)$, but it is far too difficult for the vast majority of children to understand. Yet Hart found that 33 per cent of the twelve- and thirteen-year-olds studied gave the correct answer to $3 \div 5$, $^3/_5$, on a computation paper. It is difficult to judge the processes that children used to obtain the correct answer to this question. More studies need to be conducted to determine the thought processes which children use, if indeed they use anything other than a rote method. There appears to be no evidence of children's ability to reverse the process and write $^2/_7$ as a division, $2 \div 7$. Yet with the introduction of calculators into schools, this is a piece of knowledge which is essential if fractions are to be converted to decimals using a calculator.

The fifth meaning of fraction is associated with the concept of *ratio*. A fraction is the result of comparing the number of objects in two sets or two measurements. The use of Cuisenaire rods or linking cubes illustrates this model particularly well. The length of the grey rod is three-quarters of the length of the black rod (Figure 8.21). If the parts of each rod are separated, the continuous model becomes the discrete model (Figure 8.22). The ratio meaning of a fraction is unlikely to be experienced by many primary school children, as the idea of ratio is known to be a difficult one, and is therefore frequently left until much later in the learning development of fractions.

The concept of a fraction develops over a long period of time, during which chil-

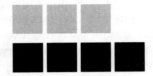

Figure 8.21   The ratio model of faction: the continuous case

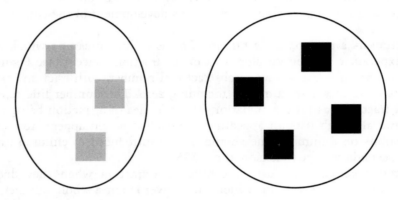

Figure 8.22   The ratio model of fraction: the discrete case

dren experience the different meanings of fractions in a variety of situations. The experiences which teachers provide will necessarily involve other mathematical concepts, including number, length, weight and money, and should be set in meaningful situations to which children can relate.

CHAPTER 9

Learning and teaching elementary algebra

THE PROBLEM OF ALGEBRA

It is not new to claim that algebra is unpopular with many children. Cockcroft (1982, p. 60), for example, claimed that 'algebra is a source of considerable confusion and negative attitudes among pupils', a conclusion arrived at on the basis of interviews with former pupils, and Herscovics (1989, p. 60) stated that 'algebra is a major stumbling block for many students in secondary school'. There are two obvious possible sources for 'confusion' and what might appear to be 'stumbling blocks', one being unhelpful teaching and the other being what is often referred to as learning difficulties. Both of these possible reasons need to be addressed. Given the existence of confusion, however, negative attitudes would certainly be likely to develop, partly as a consequence of the difficulties, but perhaps also partly because of what children perceive as the lack of contact of the subject matter with the real and the immediate. Algebra, whether experienced as a subject in its own right or whether encountered when it is applied across the mathematics curriculum, is usually presented as being essentially abstract, and seems for many pupils to be the least concrete part of what they know of as mathematics, the part which is furthest removed from their world. Algebra is first and foremost the language of mathematics, although it often emerges for pupils as a sort of generalized arithmetic, and later seems to be concerned with manipulating variables, statements and equations. On the one hand algebra enables conciseness and precision for the mathematician, because of the brevity of the symbolism, and hence of mathematical arguments and proofs. On the other hand it unfortunately often obscures, for children, the underlying meaning. Thus, although it is an essential component of mathematics, great care needs to be taken in teaching algebra, and the greatest help possible needs to be provided for pupils when they are studying this branch of mathematics.

Cockcroft also claimed that 'there has been an increase in the number of more difficult topics of an algebraic kind', a comment which was made with particular reference to the changing nature of external examinations at 16+. It was suggested that pupils from a much wider ability band were expected to cope with algebra than was the case fifty, or even twenty, years ago. This was and still is true, but more recent developments, culminating in the National Curriculum as an entitlement for all children, have led to much of the more advanced algebra, such as quadratic equations and algebraic fractions, being deferred because of the difficulties this wider ability band generally experience. This move has not been popular with many university teachers of mathematics (see Barnard and Saunders, 1994, for example), who claim they now have to remedy what they regard as vital deficiencies because of the knock-on effect at A level. Nevertheless, the majority of pupils of all ages should now be following a mathematics curriculum which encourages and enables a growing awareness and understanding of algebra, even though the demands may be different from and possibly less than they were for a minority of pupils in the past.

This does not imply that formal algebra now begins in the primary school, but it does mean that relevant mathematical activities do exist at all levels, and are important for all age groups. In recent years, considerable effort has gone into devising and suggesting alternative approaches to algebra which allow a more accessible introduction and a more gradual or different approach to formalism. Indeed, a glance at school mathematics textbooks of today and of fifty years ago reveals the enormous changes which have taken place in order to try to make algebra more meaningful for children. Furthermore, a variety of computer software which can facilitate and even promote algebraic thinking also exists.

ALGEBRA THROUGH PATTERN

One possible approach to algebra which is now widely advocated, in many countries around the world, is through the use of number patterns. According to the DES (1988, p. 16),

> Algebra develops out of the search for pattern, relationships and generalization. It is not just 'all about using letters' but can exist independently of the use of symbols. Work in the primary school on number pattern and the relationships between numbers lays the foundation for the subsequent development of algebra.

These are strong claims, seemingly without much evidence to support them, and they require a thorough consideration.

It was suggested by Küchemann (1981a) that the approach to algebra through pattern builds on relatively concrete situations which are well understood, but can be used to encourage generalization and therefore the use of letters. The example quoted by Küchemann (see Figure 9.1) illustrates very well what one might seek to achieve using pattern. The 'towers' represent different tilings with black and white tiles, and lead to data shown in Table 9.1.

The prerequisites for reaching even this stage are:

(a) recognizing or acknowledging the three relevant variables;
(b) collecting or obtaining the data;
(c) sorting the data according to variables; and
(d) ordering the data.

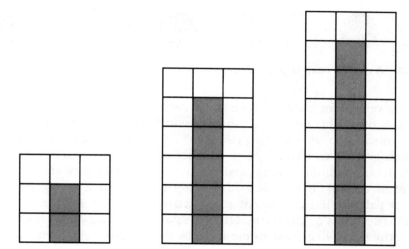

Figure 9.1   A simple tiling pattern

Table 9.1

Number of rows (x)	Number of black tiles (y)	Number of white tiles (z)
3	2	7
6	5	13
8	7	17

The algebra will then emerge only if three letters are used to represent the three variables, though, for some educators, an acceptable 'pre-algebra' stage is to use words and sentences. Thus, using the likely language of children, 'the number of black tiles is the number of rows take away one' ($y = x - 1$), 'the number of white tiles is twice the number of rows add one' ($z = 2x + 1$), and 'the number of white tiles is twice the number of black tiles add three' ($z = 2y + 3$). Teachers might suggest to the children that more data are collected, so that the numbers of black and white tiles are obtained for 1, 2, 3, 4, 5, 6, 7, 8, 9, . . . first, before the various relationships are generalized in either word or algebraic form. Some children may, however, perceive the relationships immediately, without more data. In fact, this example ably illustrates that teaching materials based on the approach to algebra through pattern usually appear to assume that symbolizing is the final stage in the process of generalizing. We cannot be certain that this assumption is either valid or correct (see Sutherland, 1990). Not only that, this 'final stage' can still be extremely difficult for children.

Patterns can be used with very young children with, for example, coloured counters, necklaces of coloured beads and lines of coloured pegs on a pegboard (see Figure 9.2). The pupils can be asked to continue the patterns of colours, or to devise their own, as part of a long-term strategy to use pattern whenever appro-

Figure 9.2   A simple bead or peg pattern

priate, as a vehicle for learning mathematics. Patterns of colours might lead next to patterns or configurations of shapes, which then might lead to number patterns, which then facilitate this particular approach to algebra. In recent years, many such situations which lead to number patterns have been suggested; for example, a variety of dot, square and triangular configurations are suggested by Mason *et al.* (1985). Examples of dot patterns are shown in Figure 9.3. The matchstick patterns (Figure 9.4) and number triangle (Figure 9.5) were adopted for research experiments by Orton and Orton (1994). The matchstick patterns lead to linear relationships, like the ones produced by the tiling patterns, whereas the dot and number triangle patterns give rise to quadratic relationships; that is,

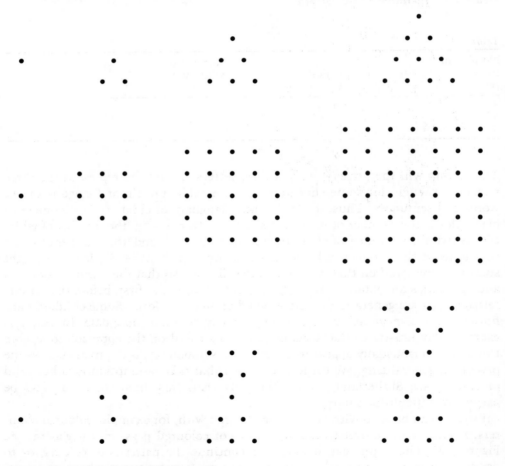

Figure 9.3   Examples of dot patterns

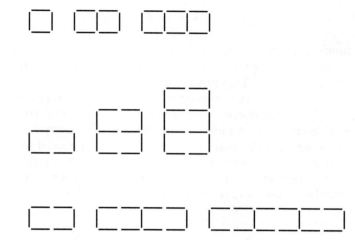

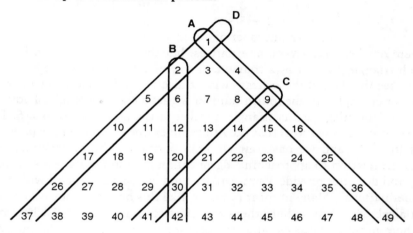

Figure 9.4 **Examples of matchstick patterns**

Figure 9.5 **A number triangle**

equations which involve a term in x^2. The reader should at this point attempt to find the 'nth' terms for the examples included in Figures 9.3 to 9.5 in order to consider which are easy and which are not.

The critical question is, of course, whether such a route to algebra helps the pupils. In fact, some educators might hope that the approach to algebra through pattern is better than any which has been used before. A considerable amount of research has been completed in recent years which, not surprisingly, indicates not that the approach will solve all our teaching problems, but what the difficulties and obstacles are when we use this particular approach! One major problem is that pupils can become fixated on what is sometimes called a 'recursive' method which does not help in the long run. This is illustrated by the pattern of triangle numbers (1, 3, 6, 10, 15, 21, . . .), which frequently arises from situations set up to produce number patterns. Here, as with many number patterns, a difference table is usually recommended in order to reveal the patterns more clearly:

$$1 \quad 3 \quad 6 \quad 10 \quad 15 \quad 21$$
$$2 \quad 3 \quad 4 \quad 5 \quad 6$$
$$1 \quad 1 \quad 1 \quad 1$$

It now becomes apparent that, because of the 'constant difference' on the bottom row, the next number on the middle row is 7, so the next number of the original number pattern is 28. This process can be repeated to produce more triangle numbers (36, 45, 55, . . .), but no matter how many more terms are derived, it is unlikely that children will obtain a formula this way. In other words, the recursive, difference table approach is a relevant mathematical activity here as it always is, but it does not automatically lead to algebra. The children need to be persuaded to look at the pattern in other ways if they are to make further progress.

In fact, progress really depends only on knowledge of numbers and their sums. The triangle numbers can be perceived as a pattern of sums of consecutive numbers:

$$
\begin{array}{ll}
1 = 1 & (n = 1) \\
3 = 1 + 2 & (n = 2) \\
6 = 1 + 2 + 3 & (n = 3) \\
10 = 1 + 2 + 3 + 4 & (n = 4)
\end{array}
$$

$(n = $ the number of numbers in the sum)

From here on there are several alternatives, but the simplest method is to represent each triangle number by a triangular array of dots, as shown in Figure 9.6. It is easy to see that two of these arrays will fit together to form a rectangular configuration; for example, for the fourth number we get a 4×5 rectangle of dots in this way, so the 'formula' for the triangle number is therefore half of 4×5. Further examples convincingly reveal that the nth number must be half of $n \times (n + 1)$, that is, $\frac{1}{2} n \, (n + 1)$, and we have our algebraic formula. The disadvantage of this approach in relation to approaching algebra through number patterns is that it is unique, and does not provide a standard method which can be used with a variety of number patterns. Many number patterns, like this one, are not easily converted into an algebraic formula, but others are relatively transparent. Teachers need to know their number patterns, and use appropriate ones to suit the purposes in mind.

Another problem with using number patterns in the approach to algebra is that pupils are often tempted to use inappropriate methods in a misguided attempt to shorten their working. The 'short-cut' method is when, for example, the twentieth term of a sequence is obtained by multiplying the fifth term by 4, or the hundredth term is taken as the twentieth multiplied by 5. This is an incorrect method for most number sequences. The 'difference product' method is when a constant multiplier, for example the number 3, is adopted and the twentieth term is then assumed to be 3×20, the hundredth is 3×100 and the nth term is $3 \times n$. Naturally, a third category of problems is that due to arithmetical incompetence. Teachers need to be alert to such possible misconceptions.

The use of patterns provides one of a number of routes to algebra. It offers some benefits, but it is not entirely unproblematic. Often children enjoy working with patterns, and, for some educators, that alone would justify the inclusion of patterning work in the mathematics curriculum. But it is important to strive for the algebra, and using patterns seems to imply introducing the use of letters,

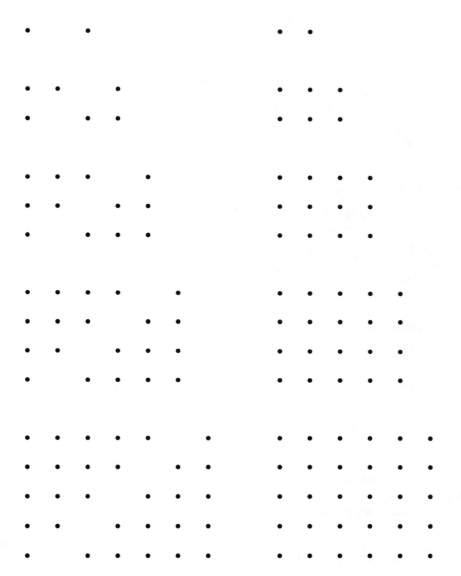

Figure 9.6   Sums of triangle numbers

leading subsequently to generalization and ultimately to the formula for the *n*th term. It is certainly not sufficient to be content with being able to continue number patterns *ad infinitum*. There is evidence from research that children can generalize, at a succession of levels, starting perhaps with methods for obtaining successive terms, leading to summing up a relationship in words, and ultimately perhaps even producing as formula of a sort (see, for example, Orton and Orton, 1994). Some children will eventually be able to carry forward their methods from one number sequence to another, thus extending their generalizing abilities still further.

SOME DIFFICULTIES IN LEARNING ELEMENTARY ALGEBRA

In mathematics, letters are used in many ways (see Wagner, 1983). Letters can be used, for example, to label points and angles, as fixed numbers (for example π), as more than one number (as in $y = x + 3$) and as units (like 'm' for metres), to list just a few possibilities. Sometimes we may refer to a letter as an 'unknown', but at other times as a 'variable', even when it is difficult for pupils to perceive any difference in what is being described. Research carried out by the CSMS (Concepts in Secondary Mathematics and Science) team (Küchemann, 1981a) revealed letters being used in very elementary algebra in six distinctly different ways:

1   The letter is assigned a numerical value from the start;
    'What can you say about a if $a + 5 = 8$?'
2   The letter need not be evaluated, and is not used directly;
    'If $a + b = 43$ then $a + b + 2 = $?'
3   The letter is used as an object, or as the shorthand for an object;
    '$2a + 5a = $?'
4   The letter is used as a specific unknown number;   'Add 4 on to $n + 5$.'
5   The letter can have more than one value;
    'What can you say about c if $c + d = 10$ and c is less than d?'
6   The letter is used as a variable; 'Which is larger, $2n$ or $n + 2$?'

There are several potential difficulties for children here. First, the very fact that we use letters in these six different ways may create confusion for children. Teachers may even switch quickly from one use to another, and may not be aware that they are doing so, and of the possible difficulties which may be created for children. Secondly, many pupils (age range 13–15) who were tested as part of the work of CSMS could not consistently cope with types 4, 5 and 6, and thus we have a clear suggestion that certain particular kinds of items are more difficult than others. In fact, Küchemann suggested that types 1–3 are so elementary, anyway, that the ability to cope with them hardly represents any real progress with algebra.

   The use of letters as objects raises another potential difficulty. In order to make the beginnings of algebra meaningful, it has become almost a tradition that teachers use 2 apples + 3 apples to give meaning to $2a + 3a$, and 2 apples + 3 bananas for $2a + 3b$. The fact that apples can be added but apples and bananas cannot certainly makes a valid point, yet the longer-term aim is to use letters as variables, in other words as numbers. Thus, strictly speaking, a and b should always represent numbers, not objects, and the short-term convenience might therefore not only serve to obscure the long-term objective, but render it less achievable, because it is never easy to persuade learners to replace one notion with another. In fact, this particular teaching strategy is likely positively to encourage the common incorrect answer, '$b + r = 90$' to the question:

   'Blue pencils cost 5 pence, red pencils cost 6 pence.
   I buy 90 pence worth.
   If b is the number of blue pencils bought and r is the number of red pencils bought, what can you write down about b and r?' (Küchemann, 1981a).

Even when this difficulty is acknowledged, and care is taken not to use letters as objects, there is a tendency for teachers to use the initial letter of the word for an object to represent the number of objects, because this aids memory, but the obvious danger is that children might subsequently misinterpret. In other words, if *s* represents the number of students, every effort must be made to ensure that the pupils do not reinterpret this as '*s* stands for students'. Some people would even suggest that a different letter should be used so as to avoid this confusion, and thus *n* might better represent the number of students.

In this context, Herscovics (1989, p. 64) reminds us of what has now become a very well-known problem:

'There are six times as many students as professors at this university. Express this fact as an equation.'

Even though one group to whom the question was originally put consisted of relatively mature college engineering students, only sixty-three per cent of them 'gave a correct answer like $S = 6P$[*sic*]'. Subsequent research with other groups of students confirms the high frequency of the incorrect answer, $6S = P$, the explanation given being that 'for each six students there is one professor'. This seems to indicate an aggregation of errors already described above. Not only does '6S' stand for 'six students' in the incorrect answer, the equals sign is also misused, and indicates correspondence, not numerical equality. This type of error, in which an equation is presented 'the wrong way round', is now widely recognized, and all teachers should be alert to the possibility of its occurrence in algebra lessons.

According to Booth (1984), children may also often 'interpret letters as standing for specific numbers, with different letters necessarily representing different values'. Another well-known problem (see Herscovics, 1989) concerns the use of equations like

$7 \times W + 22 = 109$ and $7 \times N + 22 = 109$

Here, less than half of the pupils seem to realize that *W* and *N* must have the same value, and there is a strong likelihood that pupils will have to solve both equations in order to discover this for themselves (if their arithmetic is good enough to do that!). Many experiments have revealed that some pupils believe that *W* must inevitably be larger than *N*, because it is further along the alphabet, and indeed even that the letters *A* to *Z* represent the numbers 1 to 26. It is speculative to suggest that playing with sending coded messages using numbers to disguise letters, another common classroom activity, might have influenced the beliefs of such pupils!

Booth (1984) also raised the wider issue that difficulties may not be algebraic in nature at all, but may be caused because of errors in the arithmetic. Going back to the equivalent arithmetic often does help pupils with algebra, but if they are going to commit arithmetical errors it is unlikely to be of much assistance. Also, taking the issue one step further, the arithmetical methods which pupils use are sometimes informal, even idiosyncratic, and are not conducive to setting up algebraic equivalents. If the teacher should wish to use the taught arithmetical method as a model for the algebra, it would first be necessary to get the children to accept what had been taught, which is often not easy. The CSMS Project revealed very clearly how many pupils resist teacher-taught methods and prefer their own, which are presumably more meaningful to them anyway.

There also seems to be a tendency for pupils to regard algebraic statements as somehow requiring completion. For example, '$x + 7$' might be regarded as unfinished, or 'not closed' to use the usual technical term. Collis (1975) explained that children are reluctant, to say the least, to hold unevaluated operations in suspension. Teachers at all levels will almost inevitably become aware that even two numbers connected by an operation will immediately be replaced by the result of the operation by most children, and therefore that arithmetical and algebraic simplifications seemingly must be computed at each stage, all the more so in these days of pocket calculators. Thus $3 + 4 = 7$ and $4 \times 6 = 24$; but what does that tell children about $m + n$ and $m \times n$? Does this go at least some way to explaining why $2 + x$ can become $2x$, and $7a + 2b$ can became $9ab$, in the eyes of children? When expressions like $x + 7$ and $(a + b)/(a - b)$ can be accepted as finished, closed, complete in themselves, and not being open to further abbreviation and simplification Collis described this as 'acceptance of lack of closure'. In general, he suggested that it is only round the age of fifteen years that pupils become able to accept such lack of closure.

Matz (1983) drew attention to the related problem which is created when symbols are juxtaposed in arithmetic and algebra. In arithmetic, 23 is short for $20 + 3$, but in algebra, ab is short for $a \times b$ (see Chapter 6). Although school algebra is often described as generalized arithmetic, it is very misleading that juxtaposition should imply such different things, and the difficulties created seem likely to add to the problems of reluctance to accept lack of closure. Matz also drew attention to the difficulty many pupils have in accepting that $-x$ can represent a positive number, and of course that $+x$ can be negative.

A variety of studies has revealed that many young children view the 'equals' symbols ($=$) as an instruction to do something, rather than as an indication of a relationship. Thus $5 + 3 = \square$, and even $5 + \square = 8$, are solvable, but $8 = \square + 3$ is not, and might even be considered to be 'the wrong way round'. In fact, the equals sign appears to be a more difficult symbol to come to terms with than most teachers realize (see also Chapter 6). This indicates that arithmetical equations must be adequately understood before letters are introduced. Herscovics and Kieran (1980) suggested questions like the following (with specimen answers provided for each question):

1 Can you use the equal sign with an operation on both sides?
   $5 \times 4 = 4 \times 5$
   $2 + 6 = 6 + 2$
2 Can you give an example with a different operation on each side?
   $5 + 5 = 5 \times 2$
   $3 \times 4 = 15 - 3$
3 Can you have more than one operation on each side?
   $4 \times 3 + 1 - 3 = 3 \times 2 + 4$
   $3 + 5 + 4 = 12 - 4 + 4$
4 What is the hidden number in this equation?
   $+ 5 + 4 = 12 - 4 + 4$

and subsequently

$6 + \square - 1 = 5 \times 2$

5 Only at the final stage of teaching should letters be used, in for example,

$2 \times a + 5 = 7 + 6$

It was suggested that such a teaching scheme should precede the 'think of a number' problems which are a widely used and generally popular initiation into algebra. A simple example of this approach is:

I think of a number, double it, add 3, and the answer is 11. What was the number I first thought of?

It is easy to invent such questions, and they do lead to the setting up of very simple linear equations, in this case $2x + 3 = 11$. The children, however, use guesswork ('trial and improvement', in contemporary parlance) to obtain the answers, so 'think of a number' problems therefore do not automatically lead to the usual taught method for solving linear equations. In fact, it is very difficult to get children to accept as necessary a method which involves 'subtracting the same number from both sides', and 'dividing both sides by the same number', even when the analogy of 'balance' is adopted (see Figure 9.7), because they know a quicker way to find the answer, namely to guess. Herein lies another difficulty of teaching and learning algebra.

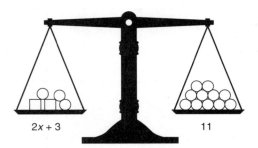

$2x + 3$　　　　　11

Figure 9.7　The balance method for solving linear equations

Figure 9.7 is a good example of a short-term aid to learning which proves unsatisfactory if it is pursued too far (see Skemp, 1971); it does not have long-term value. This particular idea illustrates some equations very well, for example:

$$x + 1 = 3$$
$$2x + 1 = 7$$
$$2x + 3 = x + 5$$

but does not really help with:

$$x + 5 = 0$$
$$x^2 = 4$$

and

$$x^2 - 3x + 4 = 0$$

Many such short-term aids exist in the teaching of algebra. Figure 9.8 illustrates the solution of the simple linear equation:

$$2x + 3 = x + 7$$

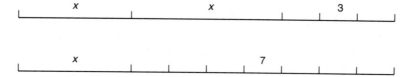

Figure 9.8   A diagrammatic method for solving linear equations

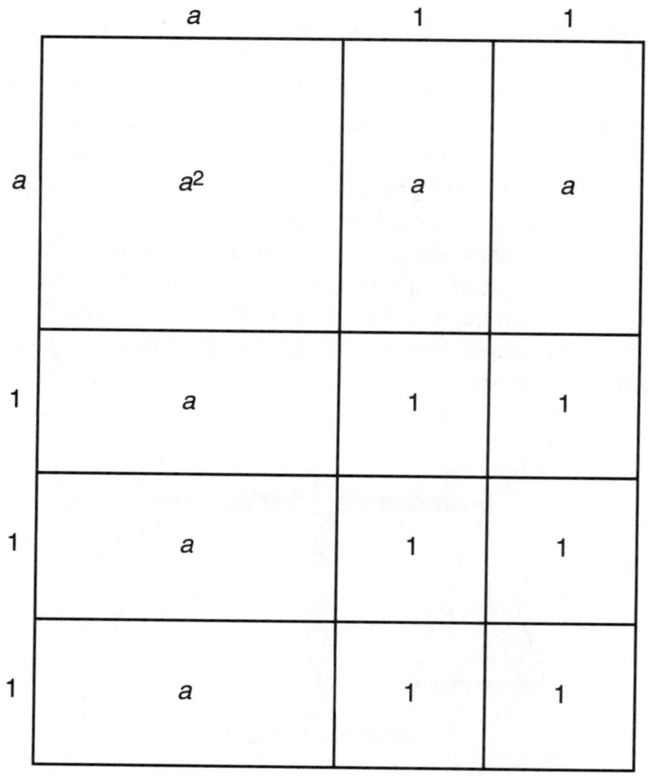

	a	1	1
a	a^2	*a*	*a*
1	*a*	1	1
1	*a*	1	1
1	*a*	1	1

Figure 9.9   A simple illustration for the distributive law

but suffers from the same problems as the above when the equations are more difficult.

Figure 9.9 illustrates a version of the distributive law, and shows that

$$(a + 2)(a + 3) = a^2 + 5a + 6$$

but it does not really help when numbers are negative as in

$$(a - 2)(a + 3) = a^2 + a - 6$$

or

$$(a - 2)(a - 3) = a^2 - 5a + 6$$

Such short-term pictorial approaches may be better than nothing, but they generally provide help only at the very beginning of a topic. There is a possibility that they could mislead some pupils in the longer run.

This is not the end of possible difficulties. For example, pupils also find sorting out the sequence in which operations must be completed troublesome, just as they do sometimes in arithmetic. Nevertheless, many of the best-known difficulties have been reviewed here. Despite paying due heed to known difficulties and problems, however, we can never expect learning algebra to become a smooth and steady progression in the development of understanding. Regression to previous incorrect notions is a persistent feature of human learning, anyway. As Driscoll (1982, p. 128) reminds us:

> Cognitive development is an ongoing process of assimilating information into conceptual schemes and adjusting the conceptual schemes accordingly. It is not always a process that progresses smoothly, however, and so occasional lapses in students' algebraic skills and understanding should be expected by teachers.

COMPUTERS AND ALGEBRA

It has become clear throughout recent years that computers offer some potential for the teaching and learning of mathematics, including elementary algebra, even though they may introduce difficulties of their own. In fact, Herscovics (1989) pays no heed to the use of computers in his review of obstacles to learning algebra. At present, there are three obvious ways of using computers which have been tried out, namely programming (usually in the programming language called BASIC), Logo and spreadsheets. Very little guidance in the use of computers in learning algebra is provided by the National Curriculum of England and Wales, though the successive versions (1989, 1991, 1995) have variously included mention or illustrations of all three possibilities. Study of these successive curricula would, in fact, seem to suggest that enthusiasm for the use of computers has waned, but this is only one possible interpretation. Another is that it has been decided that direction and guidance in the use of computers in any part of the curriculum should come from other sources, independent of a legal document which would bind teachers to using computers in particular ways. The assumption made in this chapter is that computers are here to stay in mathematics teaching, and that, if they offer possibilities of helping pupils to learn algebra, teachers need to know in what ways.

Programming

Historically, the first aspect of computing to be considered as a legitimate part of school mathematics was programming. This is hardly surprising, given the close relationship between mathematics and computer studies in the early years of using computers in schools, together with the importance of programming within early computer studies syllabuses. The second version of the Mathematics National Curriculum (1991) included the suggestion that pupils should understand the BASIC program:

```
10 FOR NUMBER = 1 TO 10
20 PRINT NUMBER*NUMBER
30 NEXT NUMBER
40 END
```

This program, when run on a suitable computer, produces the sequence 1, 4, 9, 16, 25, 49, 64, 81, 100, which at first sight just looks like another way to produce the simple, well-known number pattern of 'square numbers', particularly since it falls within the part of the curriculum devoted to 'generating sequences'. One must hope that we have not been persuaded to use a computer program simply to do that, however, and the main purpose must surely be to introduce and use a variable, in this case 'NUMBER'. Notice that the variable here is represented by a word, not by a single letter such as *x* or *n*, a point which will be returned to later. Simple BASIC programs have been incorporated into mathematics lessons in recent years by a considerable number of teachers, under the assumption that programming does have something to offer in relation to giving greater meaning to symbols, and in terms of selecting, naming, using and operating on variables. Thus, some teachers would even wish pupils to write simple BASIC programs themselves. It is not yet clear from research what is the extent to which programming can contribute to understanding algebra, though Driscoll (1982) and Tall and Thomas (1991) are amongst those who provide evidence of its value.

Logo

Logo is usually regarded as an approach to geometry, and it deserves greater recognition as another approach to algebra. In fact, Logo programming can often be used instead of BASIC, though this is not followed up here. A typical simple Logo procedure is illustrated below; it produces a square when the instruction 'SQUARE' followed by a number (the required length of side) is entered on a suitable computer:

```
TO SQUARE :SIDE
REPEAT 4 [FORWARD :SIDE RIGHT 90]
END
```

We see that a variable, SIDE, has been named, using an obvious word, and this is then operated on within a procedure, an important aspect of algebra. According to Sutherland (1990, p. 164):

> One of the difficulties with 'traditional' algebra [is] that it is not easy to find introductory problems which need the idea of variable as a problem solving tool. . . . this is not the case in the Logo programming context. Logo is a language for expressing generalities and in order to express the generality it is important to name and operate on a symbol as representing a variable.

Sutherland claims that pupils become well aware of the fact that any name will do for a variable, and that single letters are just as acceptable as words. In fact, in time, pupils do eventually choose single-letter variable names, on the grounds that it saves time to do so. They have, however, made this decision for themselves, and are thus completely at home with the use of letters either singly or in combination to express variables. This is all without any prior experience of 'pencil-and-paper' algebra. 'In the Logo context pupils are not frightened or alienated by the use of symbols.' In fact, experience with Logo does improve pupils' understanding of variables. Furthermore, Sutherland claims that pupils can accept unclosed variable-dependent expressions, like the '*x* + 7' mentioned earlier, and can subsequently operate on them. Given its value in both geometry and algebra, together with the decreased use of symbols in the mathematics curriculum overall, it would seem that Logo has a claim to be included in the schemes of work of all teachers.

Spreadsheets

Spreadsheets offer enormous potential across the whole mathematics curriculum, certainly in number work, in handling data, and some would say in algebra too. Spreadsheets were originally devised for handling data from the worlds of business and commerce, where they enable, simplify and speed up practically all the data analysis which is necessary, and even enable forecasting and predicting, for example the likely consequences of particular price changes. In terms of the number pattern approach to algebra, a simple spreadsheet might show up on the screen as:

Number	Triangle	Square
1	1	1
2	3	4
3	6	9
4	10	16
5	15	25
6	21	36
..	..	..
..	..	..

In order to set this up, however, it is necessary to refer to the cells of the spreadsheet, and to devise appropriate formulae. The 'cells' used above might be labelled by the software in a kind of coordinate system, as below:

A1	B1	C1
A2	B2	C2
A3	B3	C3
A4	B4	C4
A5	B5	C5
A6	B6	C6
..	..	..
..	..	..

Then, in this case, the entries required in the cells would simply be:

Number	Triangle	Square
1	1	A2*A2
A2+1	B2+A3	

The rest of the numbers in the table can then be obtained by the facility known as 'filling down', or 'copying down'. From the point of view of algebra, 'the spreadsheet environment can be thought of as some sort of intermediary between natural language and formal algebra' (Sutherland, 1990, p. 168). Using named variables is not required in the spreadsheet environment, which therefore offers a different kind of computer-based approach from those previously discussed. 'The environment allows pupils to test out mathematical relationships without having to take on board all the complexities of a formal language.' From the point of view of deriving number sequences (not the only way spreadsheets can provide early algebra experience), the methods used are often recursive, in that successive terms are based on previous terms, as in the case of the triangle numbers. In contrast, the square numbers can easily be expressed as a formula. Thus, finding a general term can still be difficult, and for many pupils any algebraic formulation is likely to come only in the first few elements of each column, in setting up the procedure. Nevertheless, pupils are able to express generalizations of a sort. Given the power of spreadsheets, and their likely use across the mathematics curriculum, however, it is relevant to suggest that they do offer a way of providing certain elementary but important algebraic experiences.

COORDINATES AND GRAPHS

Thought of algebra usually conjures up images of using and manipulating expressions involving single letters. It must not be overlooked that an important element of algebra is to do with coordinates and graphs, and therefore also with relationships and functions. This topic is itself too big for a comprehensive treatment, therefore only certain basic aspects will be considered. In particular functions, and 'mappings', are not considered in detail here. It is also true that a great deal of research evidence which relates to this aspect of the mathematics curriculum is now available to guide what we do in the classroom, but only a few articles are referenced here.

It might seem that coordinates are usually easily learned by pupils. Often, the game of 'Battleships', 'Treasure Hunt' maps, and street plans provide an intelligible introduction, and nowadays we must not overlook cell referencing in spreadsheets. However, children do sometimes reveal misconceptions, a very common one being that created by the very introductory activities intended to relate the ideas to the real world, namely that the numbers used along the axes in these particular activities refer to spaces and not to lines parallel to the axes. (At this stage, of course, the words 'axis' and 'axes' may not have been introduced, but when they are, care needs to be taken that 'axes' is not misunderstood through confusion with the more familiar pronunciation and meaning of 'axes'.) A better early activity which avoids this problem is that of plotting shapes, rather in the way of 'Joining the dots' or 'Dot to dot' activities in puzzle books. Houses, dogs, rockets, Christmas trees, and a multiplicity of shapes can be defined by means of pairs of coordinates plotted on squared paper, and the pupils generally enjoy joining the points to reveal the shape. They can also be encouraged to devise their own shapes, using pairs of coordinates, for setting to other pupils.

Although most pupils appear on the surface to grasp the notion of coordinate pairs and the location of the corresponding point in relation to axes in two-dimensional space, Herscovics (1989) informs us that 'there is evidence that even the notions of axis and scale are more difficult than expected'. The need to measure along axes from a common origin, the need to use zero as that common origin and not unity, the need to use a consistent scale along each axis, the need to maintain the same order in the 'ordered pairs' (coordinates), the need to maintain the same scale across both positive and negative sections of an axis, the notion that dots indicate not only position and relative order but also an actual measurement, are

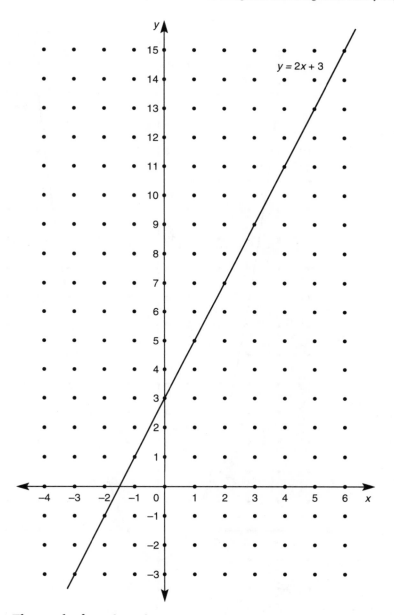

Figure 9.10   The graph of $y = 2x + 3$

all likely to take time to become assimilated. Kerslake (1981), for example, discovered that even some fifteen-year-olds were not able to provide suitable scales and position their coordinate axes correctly. It is clear that teachers must always be alert to the existence of misconceptions, even long after it might appear that the notions of axis, scale and coordinates appear to have been understood.

Most pupils do not understand the continuous nature of graphs of relationships ('functions') such as $y = 2x + 3$ (see Figure 9.10) and $y = x^2 - 1$ (see Figure 9.11), nor that the line or curve represents an infinite set of points. Many pupils find it

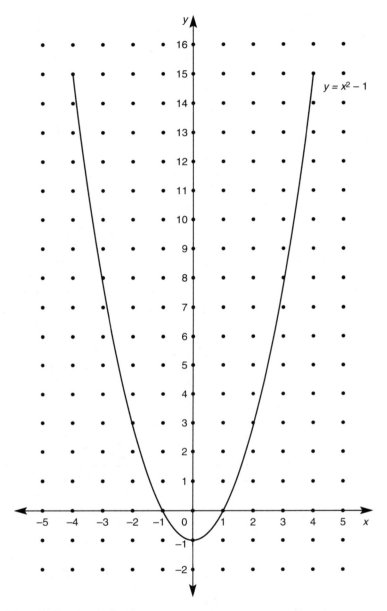

Figure 9.11 **The graph of $y = x^2 - 1$**

difficult to conceive of any points on the line or curve other than the ones they have plotted. It is no wonder that pupils are inclined to join up points even when it is not appropriate to do so, for it is likely that their understanding of the purpose and validity of joining up points is merely to highlight a pattern or shape. Teachers need to take time, when introducing the idea of joining up points, to try to justify the procedure, by plotting as many intermediate points as seems necessary to convince pupils that it is legitimate to connect them. Herscovics suggests that

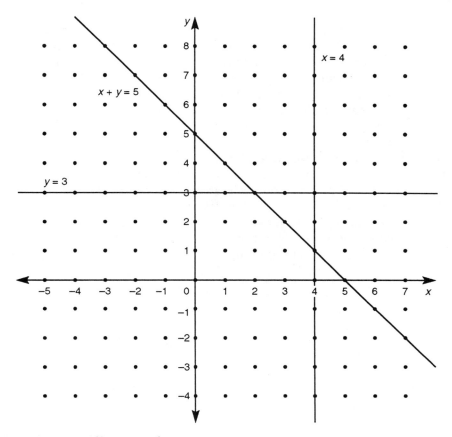

Figure 9.12   More linear graphs

the notion of covering an existing line with points is a useful intermediate step in addressing the issue of continuity. Even then, whatever teachers do, it is likely that misconceptions about the continuous nature of the lines we draw will linger.

It is also relevant to mention certain other issues of language. We are all prone to use the word 'graph' in two ways, to describe the overall picture and to describe a particular line or curve. We use the words 'horizontal' and 'vertical' to describe our axes when, for the pupils working on their desk or table, neither axis is likely to be truly vertical, and all lines are likely to be horizontal. We often describe $y = 2x + 3$, for example, as an equation for a line which we intended to graph, without realizing that we had earlier been attempting to 'solve' very similar expressions. We might appreciate the connection, but pupils might not. In other words, we need to be aware of what we are saying and doing, and we need to assist pupils to understand as fully as possible. It is also the case that the idea that a point is dimensionless, that it is a location without size, is not easy for most pupils.

Kerslake also reports a cognitive gap between being able to obtain number pairs from the equation which defines a function to be graphed, and converting these into pairs of coordinates. She also notes the difficulties experienced by pupils in identifying equations from linear graphs. Many teachers will be aware that lines parallel to the axes are often difficult for pupils, presumably because there is only one variable in each equation ($y = 3$ and $x = 4$, for example, as opposed to the more usual $y = 2x + 3$ type). Pupils also find it more difficult to produce the equations for straight lines which do not pass through the origin than for those which do, and lines of the form $x + y = n$, where n is any number, are also surprisingly difficult. These graphs are illustrated in Figure 9.12. Pupils are also weak in the interpreting of graphs, despite considerable experience with, for example, distance–time graphs. It is common for pupils to interpret lines with a positive gradient as representing 'climbing a hill'. One suggestion as regards teaching is that less time should be devoted to trivial interpretation questions which can be answered from a table of values, and more time should be devoted to focusing on intervals of a graph. Thus, we could ask about intervals over which the function represented by the equation increases or decreases, where it levels off, where it is increasing or decreasing most rapidly, where there are discontinuities, what are the rates of change over particular intervals, and what is the meaning of all these features. The search for meaning in all we expect pupils to learn must never be overlooked.

CHAPTER 10

Learning and teaching shape and space

INTRODUCTION

Our world is three-dimensional, and from birth children are surrounded by shapes of every conceivable kind. In the very early years, prior to formal schooling, children are immersed in their own movements and their positions in space relative to other objects. As children grow, the domain of conscious and unconscious experience of the world in which they live increasingly expands. Throughout their pre-school years children touch and handle many objects which have a wide variety of shapes, both natural and artificial. The aim of teaching about shapes and space in schools should be to enable children to become conscious of the variety of shapes around them, their size, how they are arranged, the way they may move and how they may change position and direction; in essence to understand their world. To achieve this aim children must be taught the skills of using their eyes to look and observe, to use their fingers and hands to feel, move, lift and turn the many shapes which are part of their everyday existence. As geometry is the science of space children should become experimenters, exploring the properties and relationships of the space which is everywhere around them. Children's innate inquisitiveness makes them natural investigators of physical objects. The teacher's role is to provide opportunities for this to occur, asking about and discussing with children the properties of shapes and how they relate to one another.

PLANE SHAPES

Our environment comprises solid three-dimensional objects which are the basic elements of geometry. Two-dimensional shapes, often referred to as 'plane shapes', have physical existence only as surfaces of solid objects. It is not possible to pick up a triangle or a square unless it is part of a solid object, such as a Toblerone

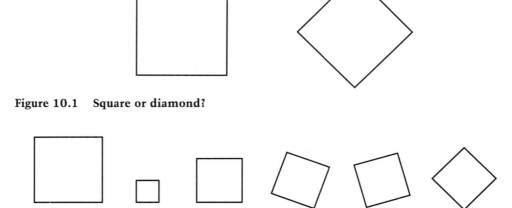

Figure 10.1 Square or diamond?

96.1	96.1	96.1	93.2	84.5	73.8

Figure 10.2 Which are squares? (The numbers are the percentage of children identifying that shape as a square)
Source: Kerslake (1979)

packet (a triangular prism) or an Oxo cube. It would thus appear that children should explore the world of geometry using three-dimensional objects, as these are part of their reality. But solid objects are complex structures and, whilst being part of a child's daily experience, are inappropriate for the study of elementary geometrical ideas and properties. Luckily, despite a wooden, plastic or paper 'circle' being in reality a cylinder (as it must have some thickness to physically exist), children readily acquiesce to this apparent anomaly, as long as the thickness of the 'circle' is very small.

Before they begin to attend school children play with shapes at home and, perhaps, in some form of pre-school education. All too often the set of shapes they use is limited to squares, rectangles, equilateral triangles and circles which only vaguely resemble the shapes they see around them. Little is known about how the very special properties of squares and rectangles influence children's development of, for example, the more general concept of quadrilateral. Many teachers will know of children who can name the shape in Figure 10.1(a) as a square, but vehemently deny that the shape in Figure 10.1(b) is also a square. Kerslake (1979) asked 103 children aged five to eleven to point to all the squares on a card on which were six squares of different sizes and orientation, positioned randomly. The percentage of children selecting each shape as a square is given in Figure 10.2.

Kerslake concludes that 'the squares were increasingly more difficult to recognize as they become more "tilted"' (1979, p. 34). She lists the percentages of children at various ages who pointed to the square which was at 45° to the horizontal. These are shown in Table 10.1. Although there were only about twenty-five children in the first three age groups the evidence suggests that around the age of eight years children begin to generalize the concept of square beyond that of the particular 'horizontal–vertical' case.

Table 10.1

Age in years	Percentage recognizing the square
5	53.8
6	56.3
7	80
8	100
9	100
10	100

In the ENCA study (Shorrocks *et al.*, 1991), 390 seven-year-olds were interviewed individually. They were given in random order a square, a rectangle, a circle, an equilateral triangle, a regular pentagon and a regular hexagon. They were allowed to handle each shape and then asked to name it. They were also asked to point to a shape which matched the one they had from a selection of pictures of the shapes on a card. The results are shown in Table 10.2. The high percentage of children matching shapes correctly suggests that they had very little difficulty with their spatial perception of shapes, in contrast to the results of the children attempting to give a shape a name. It would appear that, although young children are given experience of different shapes at home and in school, they are mainly taught the names only of squares, triangles and circles. The results are confirmed

Table 10.2

Shape	Percentage naming the shape	Percentage matching the shape
Square	96.4	97.4
Rectangle	78.1	99.2
Circle	97.4	99.5
Triangle	92.8	99.5
Hexagon	55.3	91.0
Pentagon	31.1	95.1

by a further experiment conducted with the same children. The children, interviewed individually, were presented with a set of the six shapes (square, rectangle, triangle, circle, hexagon and pentagon) which were randomly placed in front of them on a table. Each child was provided with a sheet of paper, pencil and ruler and asked to draw each shape in turn. They were told that they could use the shapes in front of them if they wished. The outcomes are listed in Table 10.3. It is significant that of the six shapes the square, triangle and circle again have the highest correct responses. What is also noticeable is that on this occasion many more children were able to reproduce rectangle, hexagon and pentagon than to give names to the shapes. Thus seven-year-old children appear to be able to identify a named shape more often than give a shape a name, indicating that when presented with shapes they could identify the named shape. The children seem to have more confidence in their ability to draw freehand squares, rectangles, circles and triangles than hexagons and pentagons. The fact that they selected the template more frequently to draw hexagons and pentagons suggests that the children were conscious of the larger number of sides of these shapes and that this

Table 10.3

Shape	Percentage correct			
	Drawing the shape freehand	Drawing the shape using the template	Method not indicated	Total
Square	76.3	15.4	5.7	97.4
Rectangle	69.9	17.0	4.7	91.6
Circle	73.6	20.6	4.7	95.9
Triangle	70.0	21.1	4.4	95.5
Hexagon	36.5	33.3	3.4	73.2
Pentagon	33.1	28.9	2.9	64.9

would prove a difficulty for them. The ENCA study also asked the children to describe a mathematical property of each of the shapes and then a further property. An average of 70 per cent were able to produce two properties for all the shapes. Ninety-two percent provided one or more properties of a square, whilst for a pentagon this reduced to 80 per cent. Overall, it would seem that seven-year-old children have a good knowledge of the basic plane shapes they have met in their early years of primary school.

The APU (undated) surveyed eleven-year-old children, looking at their responses to selecting regular and irregular concave and convex hexagons. Either nine or ten plane shapes were presented on a sheet of paper, and children had to match each shape to its name. Thus the shapes were static and the children were unable to handle them. The results are summarized in Table 10.4. The APU concluded that 'performance in matching shapes to names depends strongly on the absence or presence of plausible distractors' (p. 250).

Table 10.4

Task	Percentage matching correctly
Identify a regular hexagon with no obvious distractors present	85
Identify a regular hexagon, with pentagons present	43
Identify an irregular convex hexagon, with pentagons present	37
Identify an irregular concave hexagon, with pentagons present	25

The implications for teaching of all the findings described are obvious. Children should not be restricted to experiences with regular shapes and limited to activities involving only squares, rectangles and triangles. The names of the shapes should be used frequently both in discussion with children and in written words. 'The study of shape can promote a rich mathematical language and the vocabulary will become refined as the mathematical experiences become enriched' (DES, 1979, p. 69).

The tangram, a dissection of a square, provides excellent opportunities for children to experiment with squares, triangles and quadrilaterals of different sizes and orientations. The standard tangram has seven pieces, but dissections into three and five pieces (see Figure 10.3) are appropriate for younger children. Initially children should use the tangram with only three pieces to make their own shapes. Later they should be challenged to construct given shapes such as the one in Figure 10.4. It is important that when children have completed each construction their

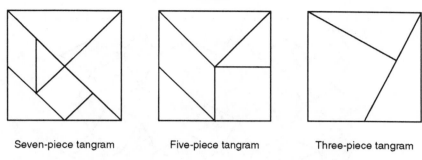

Seven-piece tangram   Five-piece tangram   Three-piece tangram

Figure 10.3   Tangrams

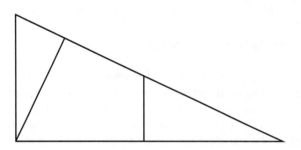

Figure 10.4   Constructing with a simple tangram

work should be talked about, using the correct name for each of the shapes, with explanations, if appropriate, of why it is so named. Thus the two triangles in Figure 10.4 should be referred to as right-angled triangles and the four-sided shape as a quadrilateral which has a right angle.

Walter (1981) describes activities for children similar to those using tangrams, but based upon pattern blocks, which are available from most educational suppliers. There are many different kinds of shape blocks now available, most of them comprising different-coloured equilateral triangles, squares, regular hexagons, isosceles trapeziums and rhombuses, and all having the same edge length (see Figure 10.5). Apparatus of this kind is a must for all classrooms, whatever the age of the children, as it has great potential for the development of many geometrical ideas, such as symmetry and tessellation (see Figure 10.6), relevant to children in primary schools and the early years of secondary school.

The principle behind the previous activities has been the construction of more elaborate composite shapes from given shapes. The reverse of this, the dissection

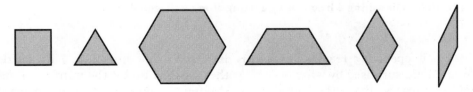

Figure 10.5   Shape blocks

137

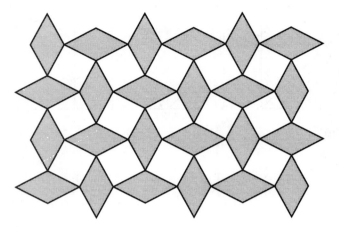

Figure 10.6   Tessellating with shape blocks

of a given shape, provides opportunities for investigative work. This will be illustrated using a regular hexagon as the shape to be dissected. An investigation which appeals to all ages relates to the drawing of diagonals. (Young children can cut or fold along diagonals if their drawing skills are not sufficiently accurate.) Teachers pose the problem, 'What different shapes can you make by drawing *one* diagonal of this regular hexagon?' Figure 10.7 shows some of the possibilities. Discussion is needed to establish that some of the dissections are essentially the same and that there are, in fact, only two different dissections using one diagonal. The natural question to ask now is, 'How many different dissections are possible using two diagonals?' Figure 10.8 shows the six different possibilities.

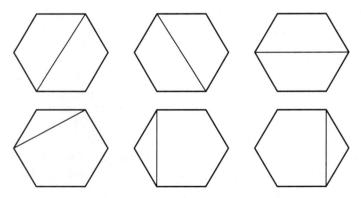

Figure 10.7   Dissecting a hexagon by cutting along a diagonal

'What happens if a regular hexagon is dissected using 3, 4, 5, 6, 7 diagonals?' Children are surprised by what occurs, as they assume that if the number of diagonals increases then the number of different possibilities also increases. Discussion on the dissection of regular hexagons necessarily demands the names

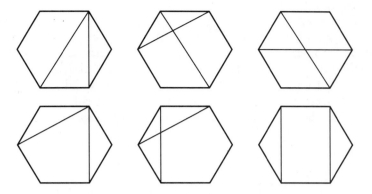

Figure 10.8   Dissecting a hexagon by cutting along two diagonals

of various shapes. It can lead to a study of the size of angles in the shapes and to the idea of congruence.

The classification of shapes is an important idea in geometry. It highlights the relationships which exist between shapes and the properties which shapes have in common as well as their differences. The DES (1979, p. 69) claims that 'classifying and discriminating are essential elements in a good mathematics programme'. The sorting of shapes into those with straight edges and those with curved edges is an early classification which young children can do as a result of observing similarities and differences between shapes. Teachers may need to construct shapes which combine the properties of straight and curved edges or sides. It is common practice in many early-years classrooms to introduce children to sorting, using Venn and Carroll diagrams, usually in relation to the colour of objects. Logic blocks are excellent for such activities. Triangles, as a special kind of polygon, result from children using a simple form of the Venn or simplified Carroll diagram, as shown in Figure 10.9.

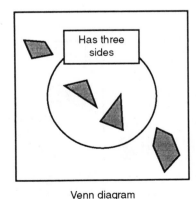

Venn diagram

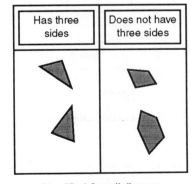

Simplified Carroll diagram

Figure 10.9   Sorting shapes

It is important that children talk with other children about what they have done, describing why shapes belong to particular 'homes'. Some children are capable of sorting shapes correctly, through their observation of whether a shape has a particular property or not, without necessarily being able to articulate their reasons. From the earliest age teachers should request that children explain to them, and to other children, 'how it works or how they did it'. To perform this kind of activity, children have consciously to reflect on the properties of shapes, matching them to the attributes specified with each 'enclosure'.

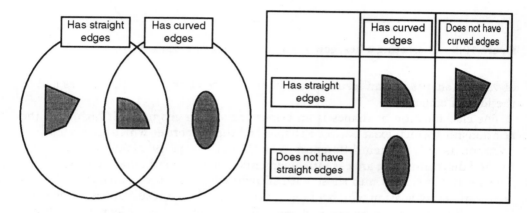

Figure 10.10   Sorting shapes using two properties

The sorting of shapes can be further extended to include two properties, demanding that children observe even more closely the attributes of the shapes to be sorted. Figure 10.10 illustrates how curved and straight-edged shapes are sorted using these two methods. Activities of this kind should only be undertaken using cut-out

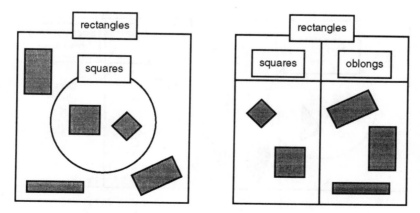

Figure 10.11   Sorting rectangles

Place these properties in their correct positions. One has been done for you.

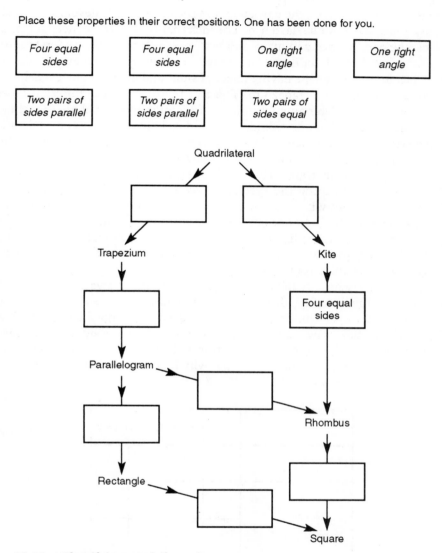

Figure 10.12   Classifying quadrilaterals

shapes which children can move around. Teachers should observe children working on such activities, as there is much to be learned about how children think when classifying shapes in this way. Kerslake (1979) discusses the difficulties children have in perceiving a square as a special kind of rectangle. The concept of set inclusion is a stumbling-block for many children, perhaps because they are seldom required consciously to perform activities which require them to use the inclusion concept. Sorting activities provide the opportunity for teachers to give children such experiences (see Figure 10.11). It cannot be over-emphasized how important it is that children should discuss 'what they did and why' when taking part in activities of this kind. Teachers may wish to relate the concept of 'oblong' to 'non-square rectangles', thus stressing that these too, like squares, are a kind of rectangle.

The relationship between the particular and the general only develops over a long period of time and cannot be hastened. The idea that the general concept of 'quadrilateral' can be particularized in many different ways is beyond the understanding of most young children. If children are asked to draw a quadrilateral the usual response is a four-sided shape having unequal sides and unequal angles. Seldom do they draw a rectangle or a trapezium. The fact that each of the different types of quadrilaterals has its own name is sufficient for children to conclude that they are not the same as quadrilaterals. Children should be helped towards an understanding of the structural relationship which exists between the different classes within a more general class, for example 'quadrilaterals' (see also Chapter 7).

The activity shown in Figure 10.12 attempts to assist children to develop the conditions which determine the special nature and name of different classes of quadrilaterals. The activity is followed by children drawing a representative of each of the classes which do not belong to any classes 'below' them on the hierarchical diagram. You might like to try this yourself. As you will have seen, this is a complex diagram and children should work with the simpler structural diagram for triangles first as an introduction to the idea.

Another activity which helps children to become more aware of the relationships between different quadrilaterals approaches the sorting idea from a contrasting perspective. Children are presented with a large matrix (see Figure 10.13), in which they draw a shape in each region, where possible, which has the necessary attributes. Drawing or constructing a shape given its properties is easier than drawing a shape given only its name, which demands that the properties of the named shape are known.

		Pairs of parallel sides		
		0	1	2
Pairs of equal sides	0			
	1			
	2			

Figure 10.13   Drawing shapes using properties of sides

REFLECTIVE AND ROTATIONAL SYMMETRY

One of the most popular activities in the primary school mathematics curriculum is symmetry. Every series of primary textbooks, together with the accompanying Teachers' Guides, devotes many pages to descriptions of appropriate activities on the symmetry of plane shapes (see Figure 10.14). Symmetry can also readily be

You need a plain piece of paper.
Fold your paper in half.
Draw a part of a tree against the fold.

Colour the part of your tree.
Cut out your tree.
Does your cut out look like this?

Your shape is symmetrical. It has a line of symmetry.

Figure 10.14   Reflective symmetry

incorporated into much larger cross-curricular or thematic topics (see Chapter 7). The APU (undated) produced a categorization of the different kinds of activities involving reflective symmetry, together with the range of facilities for eleven- and fifteen-year-old children (Table 10.5). It is apparent from these results that children find reflecting plane shapes drawn on a square grid (the APU tasks involved three- or four-sided shapes) in a 'vertical' or horizontal mirror line much easier than when the mirror line is diagonal, perhaps because the former activity is cognitively easier (J. Orton, 1989, 1994). However, it could be that as the horizontal and 'vertical' mirror line activity is frequently taught before the diagonal mirror line activity, children transfer their learning from their previous experience incorrectly to the 'more difficult' task. Thus it could be that it is the order of teaching which contributes to the order of difficulty which is so noticeable in the APU findings. Figure 10.15 shows the two tasks which the eleven-year-olds were set. Teachers would be well advised to relate the two kinds of activity early in children's experiences of reflective symmetry, drawing attention to the differences

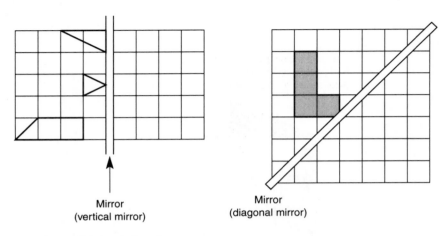

Figure 10.15   Assessing reflective symmetry

Table 10.5

Activity	Age	
	Eleven years	Fifteen years
Drawing reflections of straight-edged shapes in a 'vertical' or horizontal mirror line	80–85%	85–90%
Drawing the line of symmetry on shapes which have only one such line	50–65%	70–85%
Visualizing the effect of cutting folder paper	50%	60%
Drawing all the lines of symmetry on shapes with more than one such line	20–25%	–
Drawing reflections of straight-edged shapes in a diagonal mirror line (edges not perpendicular or parallel to the mirror line)	15%	45%

a diagonal mirror line makes to the task. The geoboard with rubber bands is a particularly helpful piece of apparatus for children exploring reflective symmetry with horizontal, 'vertical' and diagonal axes of symmetry (see Figure 10.16). It removes the demands of drawing of images, and enables children to change an attempt at a reflected shape quickly if they feel unhappy with the outcome. The image which they create can be tested using a mirror or tracing paper.

Rotational symmetry is recognized as being spatially and cognitively more difficult for children to identify than reflective symmetry. In a recent unpublished study by the authors, 507 eleven-year-old children were asked to identify which of the four patterns shown in Figure 10.17 had rotational symmetry. The children were not provided with mirrors or tracing paper and had to rely on their spatial perception and mental manipulation of the patterns to make their decisions. Forty-four per cent of the children correctly chose A, but only 33 per cent chose C. There could be a number of reasons why C had a lower facility than A. One possible explanation is that it is the only pattern of the four which has both reflective and rotational symmetry. This is an area of possible conflict which needs to be investigated further.

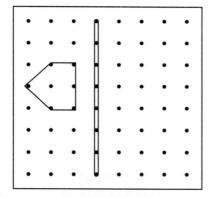 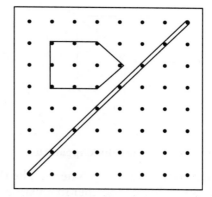

Figure 10.16   Exploring reflective symmetry with a geoboard

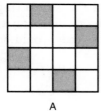 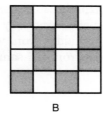 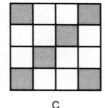 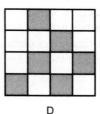

A   B   C   D

Figure 10.17   Rotational symmetry

A much fuller discussion of children's responses to reflective and rotational symmetry tasks is provided by Dickson *et al.* (1984) and Küchemann (1981b).

SOLID SHAPES

There is a language problem immediately children begin to talk about and work with solid shapes. The word 'side', previously used to signify the straight parts of the boundary of a plane shape, is now commonly used to refer to the flat surfaces (faces) of a solid shape. This confusion is unlikely to be resolved by adopting the word 'edge' for two-dimensional polygons, as 'side' is in common usage. Children can be perplexed by the double meaning of the word 'side' and must be confronted with the difficulty and not have it hidden from them.

The presentation of activities for children on solid shapes, using textbooks or worksheets, meets with the well-known problem of representing three-dimensional solids or frames on flat two-dimensional paper. Such tasks should not be given to children until they had had the opportunity to handle and construct solid objects and build three-dimensional frames. Multilink apparatus, which comprises cubes, equilateral prisms and right-angled triangular prisms (see Figure 10.18), provides children with the means of exploring many aspects of the three-

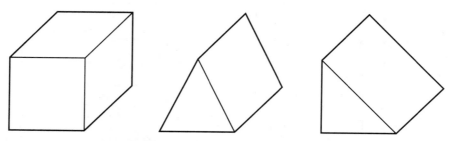

Figure 10.18   Multilink apparatus

dimensional world of geometry. The three solid shapes have common square faces and can be linked together in a variety of ways. APU (undated) looked at children's 'visualization' of two-dimensional representation of solids constructed using cubes. Children were given wooden cubes to make the solid shapes. All five of the diagrams were successfully constructed as solids by the fifteen-year-olds, and only the solid shown in Figure 10.19 created problems for the eleven-year-olds. The facility for this task was 68 per cent, perhaps because some of the supporting cubes are hidden from view. Solid shapes can be viewed only from one direction at a time and certain features of a shape may be hidden from view as they would be in the two-dimensional representation as seen from the same view.

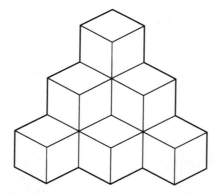

Figure 10.19   A pyramid of wooden cubes

Mitchelmore (1978) studied the styles of two-dimensional representation of a cuboid, a cylinder, a pyramid and a cube by eighty Jamaican seven- to fifteen-year-old children. He concluded that there are four stages in the developmental sequence of children's drawings of 'mathematical' solids. Stage 1 (plane schematic) drawings show a single face. Stage 2 (solid schematic) drawings show several faces, which may be visible and non-visible, but no indication of depth is apparent. Stage 3 (pre-realistic) drawings show depth from one view with only visible faces shown. Stage 4 (realistic) drawings are faithful representations showing depth and parallel edges. Typical drawings of a cuboid at each of the

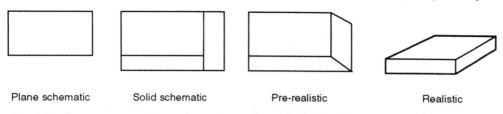

| Plane schematic | Solid schematic | Pre-realistic | Realistic |

Figure 10.20   Stages in the development of two-dimensional representations

stages are shown in Figure 10.20. If the drawings that the children make are what they perceive then it would seem that teachers must take great care in expecting children who are at Mitchelmore's first two stages to understand realistic two-dimensional representations as useful images of the respective three-dimensional solids. Children's development of two-dimensional representation can be greatly assisted if they experience making framework 'solids' using straws and pipe cleaners. The surfaces which are hidden when viewing solids are readily apparent in frameworks, although they are not surfaces in the physical sense as they do not have density and cannot be felt (see Figure 10.21). The 'solids' can be made rigid if two pipe cleaners are twisted together for each vertex and the ends to be used dipped in glue before placing into the end of a straw.

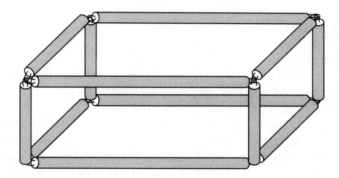

Figure 10.21   A skeletal cuboid

There are many other aspects of the shape and space curriculum which have not been discussed, such as angles, tessellations and nets of solids. Lindquist and Shulte (1987) cover much of what has been omitted in this chapter, including discussions on Logo in the classroom and the applicability of the van Hiele model of the development of children's geometric thinking. The DES (1979) sets out 'a detailed catalogue of experiences for children at different stages of development' in shape and space appropriate for teaching geometry in the primary school. This provides an excellent complement to much of the analysis in this chapter and that of Lindquist and Shulte.

CHAPTER 11

Learning and teaching data handling

THE RELEVANCE OF PROBABILITY AND STATISTICS

The area of mathematics described as 'data handling' within the Mathematics National Curriculum of England and Wales is perhaps better known to some readers as 'probability and statistics'. According to Holmes (1980, p. 15), however, statistics is 'not a subset of mathematics', it is 'a practical subject devoted to the obtaining and processing of data with a view to making statements which often extend beyond the data'. This view might reflect an attempt to justify statistics as a separate curriculum subject, or it might reflect something of a purist attitude, or it might reflect the many uses of statistics across the whole school curriculum, but it is not a view with which most mathematics teachers and curriculum planners would agree. Statistics is currently considered an important element of mathematics curricula in many parts of the world, though it was hardly taught at all fifty years ago. Probability is an inevitable companion of statistics, because models based on probability underlie most statistical theory and, to the learner, probability often appears to be very mathematical indeed. Thus, if only by association, statistics must be considered to be mathematics too.

Early ideas of probability begin with intuitive notions of chance and likelihood, and at this level probability features in everyday conversation, much more so than the rest of mathematics, and certainly more than algebra and geometry. Many of the things we do and say in our everyday lives incorporate notions of the possible and the probable; indeed, even many statements in this book do. When we wonder whether it will rain this afternoon, or whether a particular child will be absent tomorrow, or whether we shall win the lottery this week, or whether we shall have a good holiday this year, or whether we shall live to a ripe old age, we are attempting to estimate what is the likelihood of a particular event happening or not happening. Often, we are considering the chances of success or failure in relation to an event. Such interactions with daily life provide ample motivation for the study of probability, and provide us with a multiplicity of examples for mathematical study at all levels of the curriculum. It might be assumed that probability is relatively simple because of all this, but the reality is far different. The objective of teaching probability must, of course, be to enable learners to come to a better understanding of how chance and uncertainty might be measured and understood.

The word 'statistics' is used in two ways, even in schools. It is the name given to the subject or curriculum area, and it is also used to describe the very numerical products and outcomes of study and analysis within the subject (like the various averages – mean, median and mode). In order to avoid confusion, the word will be used here to describe only the subject.

Statistics is often justified within the school curriculum because it has become such an obvious part of everyday life. Numerical information in one form or another assails us from all directions, for example from newspapers, radio and television, cinema and advertising displays. Sometimes data are presented in the form of graphs which need to be interpreted, sometimes data are tabulated, and sometimes data are merely stated, but always the listener or reader needs to think about what is being claimed, and needs to think about the way in which the data have been collected, analysed and presented. Statistics has undeservedly acquired something of a poor reputation; indeed, it is even said that one can 'prove' whatever one wants to prove by quoting carefully selected data. A major objective must therefore be that children learn how to interpret data, and not always to accept them at face value, because data *are* sometimes unscrupulously misused and abused in order to try to prove a particular point. Fortunately, elementary statistics is both accessible and motivating, even for young children, because its basis can be the collection of data which are both relevant and meaningful. According to Cockcroft (1982, p. 234), 'statistics is essentially a practical subject and its study should be based on the collection of data, wherever possible by pupils themselves'.

GRAPHICAL REPRESENTATION

The collection and subsequent representation of data in pictorial form are important elements of the school curriculum even for young children. In fact, pictorial and graphical representations of various kinds constitute a very large part of elementary school statistics, yet comparatively little research has been carried out into learning difficulties in this area. Children usually enjoy drawing statistical graphs, perhaps partly because practical activity is involved and partly because the end product can be pleasing to the eye. Enjoyment is obviously important in education, but the ultimate purpose of learning to produce statistical graphs is to learn another way of conveying reliable information. Often, data presented graphically are easier to comprehend than if they were described in words or even tabulated. Sometimes, a graph might even enable us to perceive connections and relationships which are obscured by the quantity of numbers in the original data. In other words, graphs are an important means of communication, and we should obviously be aiming to enable children to communicate clearly, accurately and attractively, through graphs. We should also be aiming to ensure that children understand how graphs are sometimes wrongly used, perhaps with deliberate intent. There are many different kinds of statistical graph, for example bar

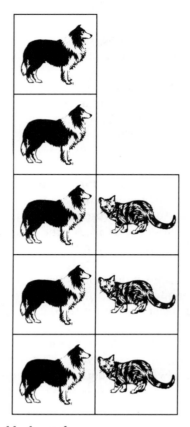

Figure 11.1 A very simple block graph

charts, pie charts, line graphs of various types, scatter graphs and histograms, and each is most appropriate for different kinds of data. They should therefore not be used unthinkingly just because, for example, the class needs to revise bar charts! We shall be able to deal here only with the basics of these common types of graph.

After the problem to be studied has been defined, essential first steps in the process include deciding what data ought to be collected and why, and then deciding how to do the collecting. In teaching, however, these important decisions will need to be dealt with not just in the early years, but throughout the whole of schooling, whenever graphs of data are studied.

The beginnings of the actual drawing itself are likely to lie in what Nuffield (1967b) and many others before and since have called 'block graphs'. Figure 11.1 is a very simple block graph which shows how many dogs and cats a small group of children have at their homes. The complexity of the task is deliberately kept at a low level in this example. Indeed, to ease conceptual understanding, plastic shapes or cut-outs of animals should be used and perhaps stuck on to a large piece of paper at this stage, in preference to any more sophisticated approach. Once this graph has been completed it is easy to develop the idea over a period of time, perhaps by first incorporating data from more children, and eventually from the

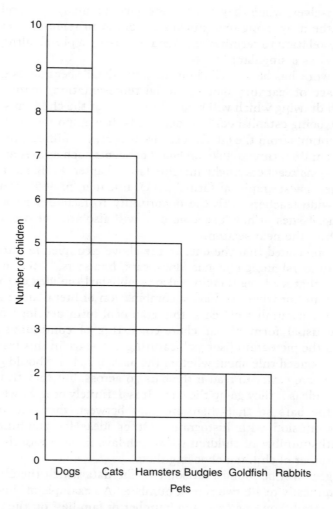

Figure 11.2   A more complex block graph

whole class. It is then a relatively small mental step for the children to extend the number of pets so as gradually to include hamsters, budgies, goldfish, rabbits, and whatever else the children might have as pets.

At this early stage there should not be an explicit 'scale', with written numbers, and even the idea of axis is still implicit. If it is intended to compare numbers of pets, however, it is necessary that the children come to appreciate that aligning the shapes or pictures is necessary to facilitate at-a-glance comparison, and this can be achieved by using paper with a grid of large squares. Eventually, axes can be drawn (though it might not yet be appropriate to use the words 'axis' and 'axes') and labelled 'pets' and 'number', and the numbers can be written along the appropriate axis (see Figure 11.2).

Finally, a scale which is not one-to-one would complete the introduction of basic ideas, but it is again best if the pupils can be guided, if necessary, to propose the idea themselves, when they realize they have too many data and perhaps need to condense them by using one unit on the graph to represent several children. (Note that, in relation to terminology, 'data' is strictly a plural, although it is often used nowadays as a singular.)

This early work has been spelled out in some detail, because even here, at this very early stage of learning about pictorial representation, many of the ground rules of graph drawing which will be used throughout the children's mathematical education are being established. In other words, it is important not only to inculcate correct notions from the outset (because it is always difficult to 'unlearn'), but also to allow children time to take on board concepts such as axis and scale, which are not as easy as teachers might imagine (see Chapter 9). 'Block graphs and bar charts are the easiest graphical forms' (APU, undated, p. 388), so these kinds of graph do provide teachers with the opportunity to focus on basic principles of graph drawing. Issues of how to record data will also arise before long, but these are dealt with in the next section.

It will be appreciated that the data in the above example are 'literal' (involving words, not numbers) along one axis (dogs, cats, hamsters, . . .) and numerical on the other, and that we have therefore also established the idea of the bar chart as being appropriate for when we have a combination of literal and numerical data. Eventually, over a number of years, this graphical form develops from the block graph to the usual form of bar chart consisting of continuous bars without pictures, with the pictogram (isotype) featuring as a stage in this transition. There is no generally agreed rule about whether the bars of a chart should go across or up the page. They are easier to read if the bars go across, but they fit better into the progression of ideas if they go up the page. It is definitely best, however, to encourage placing the bars so that there are gaps between them, in order to avoid subsequent confusion with histograms. Other ideas for graphing in this way include months/number of children with birthdays in each month, and favourite TV shows/number of children choosing them.

Before long, it will be found that some of the data which the children wish to represent graphically relate two sets of numbers. An example of this is 'number of children in a family' on one axis, and 'number of families' on the other, and this is illustrated in Figure 11.3 (the teacher needs to be aware that some children have difficulty with themselves, in that they often overlook that they too are one of the children). Other examples include shoe size/number of children, and bedtime (to nearest quarter hour, say)/number of children. (Pocket-money/number of children is guaranteed to lead to a disintegration of the lesson!) The type of graph drawn here is really a natural progression from the previous bar charts, so some people would refer to it as a 'bar line chart' or even just 'line graph', or 'vertical (!) line graph'. It is now the case that width for the 'bars' is less appropriate, and might even mislead, although it would perhaps be going too far to claim that a bar chart would be utterly wrong for these data. Each number on the 'number of children in a family' axis has a single specific point location, therefore the 'frequency' is represented by a line through that point. The technical term 'frequency' has now been introduced, but children and teachers might prefer to continue to use 'number', or even 'how many?' throughout the earlier years.

The pie chart is a temptingly attractive form of representation, but needs to be

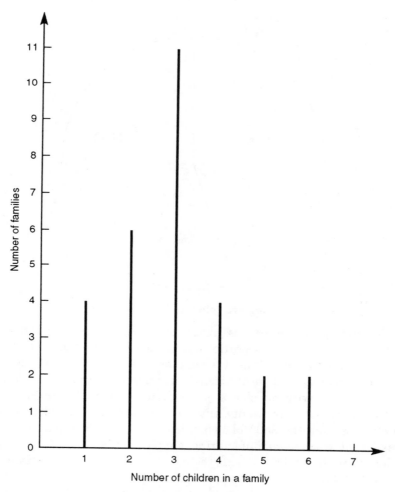

Figure 11.3 A line graph of family sizes

handled with great care. According to the APU (undated, p. 388), 'Pie charts are the most difficult graphical form . . .'. Data like eye colour/number of children could be represented very effectively in this way (see Figure 11.4). Clearly, we again have here a combination of literal and numerical data, so we could use a bar chart, but the difference is now that we are essentially dealing with proportions of a recognizable, sensible whole, namely the class of twenty-nine children. One can claim that there are only a limited number of categories (say blue, brown, green and grey) into which it is possible to place all children, and the pie chart then clearly reveals the proportions of children with each eye colour. If there are a large number of categories in which to place the children a pie chart would have too many, relatively narrow, sectors for it to be attractive and easy to read. This could happen, for example, if we collect data on each child's favourite TV show. There would be a different problem with pets, in that many children have more than one

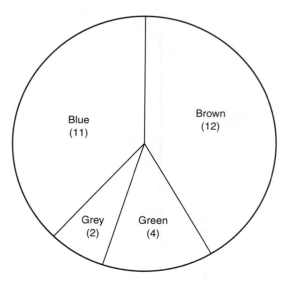

Figure 11.4 A pie chart showing eye colour of children in the class

pet, so the total number of pets is greater than the number of children, and the sectors would not represent proportions of the class of children. In this case a bar chart would be appropriate. Thus, it should be clear that, whilst there may *sometimes* be a choice of graphical form, often one form is more appropriate than another.

The real difficulty with pie charts lies in drawing them. Figure 11.4 represents the proportions of a class of twenty-nine children, and 29 is an extremely difficult number to divide into the 360° of a complete circle! Often, children are protected from this by the teacher 'manufacturing' simple numbers but, if we wish to use data collected by the children about themselves, we have to be prepared to cope with the consequences. Proportion is an extremely difficult concept for many children. Pie chart scales, which look somewhat like protractors, can be purchased nowadays, and these do simplify the drawing, but they may not be appropriate for young children. Other simple ideas for pie charts include colour of hair/number of children, and means of travelling to school (walk, bus, car, cycle)/number of children. 'How I spend my schooldays' (numbers of hours spent sleeping, eating, playing, working and watching TV, for example) is generally a popular form of pie chart, though it must be realized that this example is slightly different from the others mentioned above because it does not incorporate the true notion of frequency. A simpler version of this kind of 'time' pie chart, where the children start with a clock face divided into twelve hours, can of course be used as a simple introduction to the shape and appearance of pie charts.

So far, all our data have been 'discrete' rather than 'continuous'. Pets need to be classified discretely as, say, dogs or cats. There is no 'in-between' category ('cogs' or 'dats'?), let alone a gradual continuous transition from dogs to cats. Likewise, we usually choose to define discrete categories for eye and hair colour, as a convenience for classification and graphing, even though there are many shades of blue,

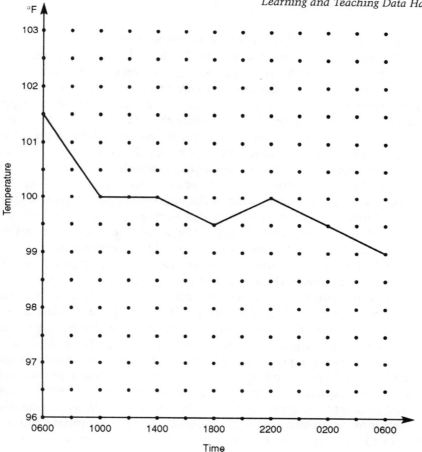

Figure 11.5 A continuous line graph showing temperature of a hospital patient

green and brown. If we were concerned with representing how the outside temperature was changing throughout the twenty-four hours of a particular day, however, we would have a situation in which the data could change continuously along both axes. Another familiar example is that of recording the continuously changing temperature of a hospital patient (see Figure 11.5). What is interesting, however, is that temperatures in both cases are likely to be noted at discrete intervals, hourly, four-hourly, or whatever, and continuity has to be assumed for the purposes of graphing. So is it more appropriate for points which are plotted on the basis of such data to be joined up by a curve or by straight-line segments? There is no categorical answer to this question, but often, as in the case of hospital patients, straight lines are used, presumably on the ground that no one can say that any particular curve is more likely than any other, and lines are easier to draw. With the kind of (x, y) graphs discussed in Chapter 9 we can be absolutely sure how to join up the points on the basis of the particular form of the equation which connects x and y. With continuous statistical graphs based on data which is not continuously collected, we are hardly ever likely to be as sure as that.

Although such 'jagged line graphs' (some people would call these 'line graphs', which can create confusion with the earlier graphical form) are very familiar, and

155

are relatively easy to draw, they must not be used loosely or inappropriately. It is helpful at this stage to introduce the notions of independence and dependence of the 'variables' represented on the two axes of a graph (axes are not relevant to pie charts), and perhaps to the tabulation of corresponding values of the variables. Often, one variable is dependent on the other; that is, one set of data is dependent on the categories we have defined for the other set. The numerical data which we would need to collect about eye colour, for example, are dependent on the colours we have chosen. If we change the colours by introducing the new category of 'hazel' we change the distribution of numbers across the categories. Yet we are not free to change the numbers for any other reason. If we are collecting temperatures, we need to relate these to some fixed scale, such as the hours of the day, otherwise they have no meaning. Thus 'eye colour' and 'time' can be classed as 'independent', and the other variable is then 'dependent'. When 'time' is used, it is almost invariably the independent variable, whilst 'frequency' is usually the dependent variable. The independent data are by convention located on the axis which goes across the page, and the dependent data lie on the axis which runs up and down the page.

Sometimes, data which ought to be represented by a bar chart or line chart are mistakenly represented by a jagged line graph. If the independent variable was, for example, 'the number of items purchased' (pens, say), and the dependent variable was 'cost', it would be wrong to connect up points (see Figure 11.6). To do so clearly suggests that intermediate points have meaning, that it is legitimate to think of purchasing 1.5 or 2.37 pens, and that the corresponding costs also provide valuable information. This particular example is an illustration of bad practice, yet it is encountered from time to time, even in textbooks. Sometimes it is done with the best of intentions, to highlight a trend, and might even be legitimate, but it is not legitimate in this example. In the long term such practices can seriously mislead children, who are all too prone to join up points unthinkingly. Strictly speaking, only when intermediate points do have meaning should points be connected, and teachers need to be alert to misuses which might promote bad habits.

Another simple graph which may be based on continuous data on one or both axes is the 'scatter graph' or 'scattergram', illustrated in Figure 11.7. In this example, data have been collected which relate children's weights to their heights, which in effect provides us with pairs of coordinates to plot (see Chapter 9). (Some children are very sensitive about their personal characteristics, so data like this need to be handled carefully.) Although we might wish to use such a graph to investigate relationship or dependency (the conjecture that taller people weigh more, for example), and even to forecast weights for given heights (or vice versa) within the range for which we have data, it is not worth making an issue over which variable is independent and which is dependent at these early stages of the study of statistics. If there were an exact relationship between height and weight, the points would lie on a straight line, of course, but this is not likely to happen. Having plotted points, however, we certainly need to discuss what we can deduce about the relationship between the variables, for a variety of scatter graphs, in order to learn how the closeness or otherwise of a relationship is likely to be revealed by the graph. Any suggestion that points should be joined needs to be discussed in order to try to remove misconceptions. It was Kerslake (1981) who first recorded research evidence that pupils may well want to join up a small scat-

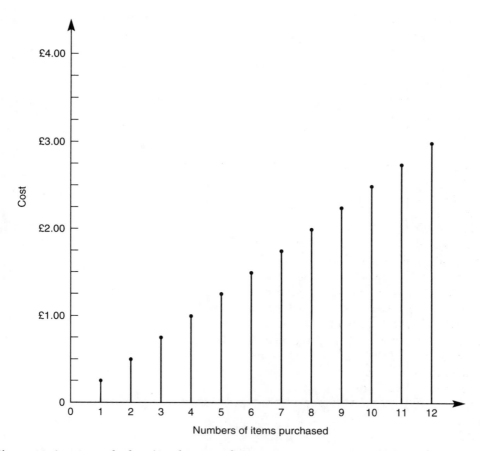

Figure 11.6   A graph showing the cost of pens

tering of points, and for a whole variety of irrelevant reasons. A large scattering of points is better for introducing the idea of 'line of best fit', which is the only line that may possibly be appropriate here. A line of best fit is usually drawn by eye in elementary statistics, though it is possible to improve validity by first plotting the point whose coordinates are the means of the two sets of values of the variables, in this case (mean height, mean weight), and then drawing the line through this point. Means are dealt with later in this chapter (see 'average'). Other examples of scatter graphs which should interest the children include such personal comparisons as height/arm span, and hand span/shoe size.

The last of the major forms of statistical graph considered in this chapter is the 'histogram' (see Figure 11.8 and Table 11.2). Here, the independent variable is continuous, but the data have been gathered into numerical categories of equal width (they do not have to be of equal width, but there are additional problems if they are not, in that it is the areas of the blocks which represent the frequencies). In other words, the data have been 'grouped' into 'classes', because this is often the only way of dealing with large quantities of data. Naturally, some of the detailed information contained within the original data is lost in a histogram,

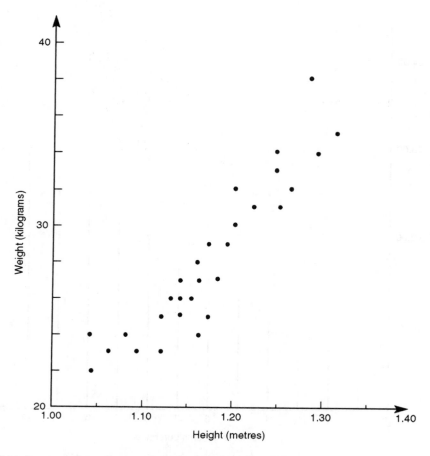

Figure 11.7   A scatter graph showing relationship between heights and weights

because of the grouping. The dependent variable represents the frequency, and is therefore discrete. At first glance, this graph might look like a bar chart with the bars placed immediately adjacent to each other, but the two graphical forms are fundamentally different and must not be confused. Again, it is possible to find textbooks in which bar charts are referred to as histograms (and vice versa), as if their superficial similarity of appearance is what determines their name. Indeed, it is essential that there are no gaps between the columns in a histogram because of the continuous nature of the independent variable. On the independent axis, numerical values can most conveniently be indicated by labelling the boundary points between classes. This is often not an easy form of graph to draw, and there are many potential minor difficulties to cope with, and sometimes awkward decisions to make. It is assumed here that the serious study of histograms is beyond the scope of elementary statistics, and that they will not be studied by children until the later years of schooling.

Whatever graphs are being studied, the children themselves should be actively involved in collecting real-world data which are meaningful to them. Also, according to Curcio (1987, p. 391):

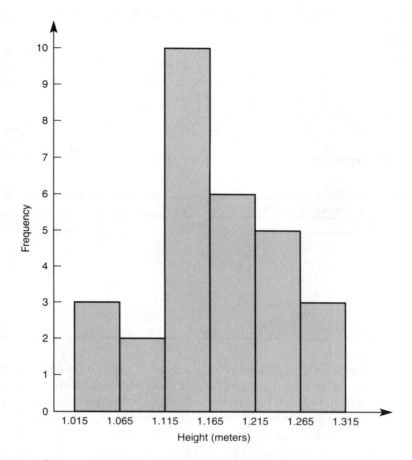

Figure 11.8 A histogram showing distribution of heights

[Children] should be encouraged to verbalize the relationships and patterns observed among the collected data (e.g. larger than, twice as big as, continuously increasing). In this way, the application of mathematics to the real world might enhance [their] concept development and build and expand the relevant [conceptual structures] they need to comprehend the implicit mathematical relationships expressed in graphs.

TABULATING DATA

In the early years, data may be collected and displayed without much attention to a systematic method. In later years, however, particularly when comparatively large amounts of data are involved,the easiest way to collect and record data is to use the 'tally' method. This is illustrated in Tables 11.1 and 11.2. As each item is obtained, a tally mark is placed in the appropriate row of the table, and each fifth tally mark is drawn across the previous four (obliquely, by custom) in order to assist in totalling. The numerical data concerning the independent and dependent variables then define what is known as a 'frequency distribution'.

Table 11.1 The number of brothers and sisters of children in a class

Number of brothers and sisters	Tally	Frequency
0	1111	4
1	1111 1	6
2	1111 1111 1	11
3	1111	4
4	11	2
5	11	2
	Total	29

Table 11.2 The heights of a class of children measured to the nearest centimetre and grouped in classes

Height (m)	Tally	Frequency
1.02–1.06	111	3
1.07–1.11	11	2
1.12–1.16	1111 1111	10
1.17–1.21	1111 1	6
1.22–1.26	1111	5
1.27–1.31	111	3
	Total	29

Note that a decision always has to be made as to how to group data, and this involves deciding what size the classes should be (which is related to how many classes) and where to begin. The nature of the data might provide some guidance, but often these decisions are at least partly arbitrary. A very large number of narrowly defined classes is probably not compatible with a decision to group, and a very small number of wide classes probably conceals too much information. And, whatever decisions are made, different groupings will inevitably lead to slightly different shapes of graphs.

AVERAGE

'Average' is a word which is frequently used in the world outside the classroom, probably often in a fairly vague way, perhaps in the sense of some kind of normal value. The fact that the rainfall in January 1995 was twice the average for that month certainly confirmed suspicions that quite a lot of rain fell, and may suggest that nature dealt somewhat harshly with us all, but many people's understanding of the concept of average may not go much further. It is unlikely to extend to the fact that one consequence is that the average January rainfall is now higher than it was, simply because of the 1995 figure. In the same way, any nationwide exhortation to teachers to produce more children who are gaining better than average marks would be ridiculous, because the average goes up as the marks improve, and always stays somewhere 'in the middle'. Furthermore, a period of very wet weather may lead to the assumption that we must be in for a dry spell, because of the so-

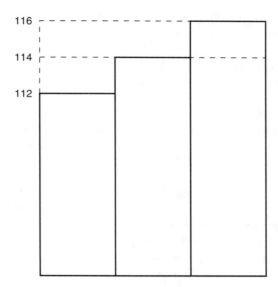

Figure 11.9 A method of calculating the mean

called 'law of averages', but events could prove that to be just as ridiculous.

The concept of average is an important one for children to come to terms with, simply because of its use in employment and society at large. In fact, there are three common measures of average, known as the 'mean', the 'median' and the 'mode', all calculated differently and all appropriate to different situations. The median is quite simply the middle value of a set of numbers, but neither of the others need be 'in the middle'. Thus the median height of pupils in a class might be the most useful average when considering the positioning of noticeboards or coatpegs. The mode is the value which occurs most frequently, and is of particular importance to manufacturers who, knowing the modal dress or jacket size, are then able to produce the right proportion of their products in that size. Neither the median nor the mode requires much calculation at the school level, once data have been collected, but it is important to devote time to discussing which kind of average is most appropriate for particular circumstances. The mean is the most likely average to be used, and would certainly be what is used to calculate the average rainfall. It is the mean for which a method of calculation is usually taught.

Suppose there are three children in the class whose heights are 112 centimetres (cm), 114 cm and 116 cm. For three numbers which are all different the mode is inappropriate, because it does not make sense to talk of the most frequently occurring value. There is a middle value, however, so the median is 114 cm. Finding the mean can perhaps best be illustrated pictorially (see Figure 11.9). Is it possible to remove a piece from the tallest column and use it to build up the others so that all three columns are the same? Yes, in this case we need only to remove the top 2 cm from the tallest and place it on top of the shortest. The mean is therefore 114 cm, which happens to be the same as the median.

If we now have three other children whose heights are 112 cm, 114 cm and 122 cm, the problem is more difficult, but can still be solved. If the top 6 cm is removed from the tallest column, and split into 4 cm and 2 cm, these can be placed on the shortest and the middle columns, respectively, thus making them all 116 cm. Thus the mean is 116 cm, which is not the same as the median. This is more difficult than the first example, but it raises the issue of what the mean is all about. There is a kind of feeling of 'fairness', or 'making them all the same', in this approach to the mean.

This idea of fairness is even more evident if three children are each given 3, 5 and 7 sweets, respectively. They will soon complain that it isn't fair, and that they should all have been given 5 sweets. How would these children sort out what was fair if they were given 3, 5 and 10 sweets, respectively? One obvious way is to pool all the sweets and then share them out equally, resulting in (3 + 5 + 10)/3; that is, 6 sweets each. This introduces the calculation of the mean in a simple yet motivating way, and again the mean is not the same as the median. The emerging definition for the mean is that it is the sum of all the numbers divided by the number of numbers. If four children are given, respectively, 3, 3, 6 and 8 sweets, the mean ('fair') number is (3 + 3 + 6 + 8)/4, that is, 5. It is very easy to make up such examples by starting with a number of children, say 5, then a number of sweets, which must be a multiple of 5 (certainly at the start), say 30, and then inventing 5 numbers which add up to 30, say 2, 4, 7, 8 and 9.

The idea can be extended to incorporate the idea of frequency, which is important in the longer term. If one child is given 8 sweets, two other children are each given 5 sweets, and three other children are each given 6 sweets, the frequency distribution is:

Sweets	Children
5	2
6	3
8	1
Total	6

Now, we first need to find out how many sweets have been given out altogether, and this consists of 5 sweets to two children, 6 sweets to three children, and 8 sweets to one child, in other words, $5 \times 2 + 6 \times 3 + 8 \times 1 = 36$ sweets. Then we need to share out these 36 sweets among the six children, so they should all have 6 sweets each, indicating that 6 is the mean number of sweets. The method of calculating the mean for the frequency distribution is therefore:

$$\frac{5 \times 2 + 6 \times 3 + 8 \times 1}{6}$$

We have now come a long way, and have taken the idea of mean across many years of schooling, but we have done so by tackling increasingly difficult but fairly realistic situations. We have also reached the position of being able to calculate the mean number of brothers and sisters from Table 11.1:

$$\text{Mean} = \frac{0 \times 4 + 1 \times 6 + 2 \times 11 + 3 \times 4 + 4 \times 2 + 5 \times 2}{29}$$
$$= 2$$

It will be realized that this is artificial and does not equate with the national average of around 2.3 (and this is liable to change over time, just like the average January rainfall), but the numbers have been carefully chosen so as to simplify the working. This in itself raises an important issue for all those who teach mathematics, for which there is no simple prescription. When is it right to simplify the mathematics with the resulting accusation of artificiality, and when is it right to expose children to realistic data and have to face the consequences?

Children also need to have the opportunity to discuss the idea that the mean is a single number which in a certain sense represents and therefore can be used to replace all the data from which it has been computed. On its own, it tells us something about the data which might not have been obvious before, whilst at the same time concealing all the detail. Thus, if only a mean is available, and no further detail about a distribution is known, pupils need to consider what, if anything, can be deduced about the distribution. In fact, it tells us that the original values must be clustered around this mean, but it does not tell us how widely spread the values might be. It requires a different and more advanced kind of statistical measurement to do that, and again there are a number of possible measures, but they are part of the mathematics curriculum for only a small proportion of students. Although, in comparison with some other branches of mathematics, there is only a limited amount of published research on children's difficulties in learning statistical concepts (see Garfield and Ahlgren, 1988), there is enough to know that teachers need to do more to help pupils to understand concepts, in addition to helping them to carry out computations accurately. Since the various averages come early in the study of statistics, and since the mean in particular is central (in more ways than one!) to subsequent statistical concepts, it is worth while making a special effort with such basic ideas.

ELEMENTARY PROBABILITY

The National Curriculum of England and Wales (DES, 1991) suggests that probability begins with the consideration of degrees of uncertainty about the outcomes of events, and this approach is widely advocated in textbooks and teaching materials. Some events are considered to be certain, or almost certain:

- the sun will rise at dawn tomorrow
- the tide will come in
- it will rain this year
- next Christmas Day will be in December
- we shall all be older tomorrow;

some are considered impossible, or virtually impossible:

- I shall grow 10 centimetres in height overnight
- my teddy will walk across the ceiling
- you will have two birthdays this year
- I will see a cow with six legs today
- all schools will soon be closed for ever;

and yet others are somewhere in between:

- it will not rain tomorrow

- I will see a red car today
- it will be very cold next Christmas Day
- I shall be a millionaire next week
- you will watch television tonight.

Children should be allowed to invent their own examples of these three categories. Thus hypothetical events can be placed in order of likelihood, on a kind of scale.

The introduction of a scale is extremely important, but it can be gradual. At first, children need only to be asked to think of:

certain	impossible

but finer definition might then lead to:

certain	possible	impossible

and then to:

| certain | likely | unlikely | impossible |

and:

| certain | very likely | likely | unlikely | very unlikely | impossible |

(The change from the suggestion of distinct categories, represented by boxes, to a continuous scale, represented by a line, needs to take place as the children's appreciation of the notion of chance and probability becomes better developed.)

As well as providing their own ideas, children can sort cards, which have events written or pictured on them, into the above categories, gradually refining and justifying their decisions. The events must always be within their own experience, of course. Young children will make their decisions on the basis of experience, but heavily influenced by prejudice and preferences, with the last perhaps dominant. There may even be conflict between these three influences, but there is little point in teachers trying to explain to young children why they are not correct in their decisions, though some discussion may be possible.

An interesting point which certainly needs to be dealt with concerns the language of probability, namely that the opposite of 'certain' is not 'uncertain', it is 'impossible', and the opposite of 'impossible' is not 'possible', it is 'certain'. Probability eventually demands a scale from 0 to 1, 0 representing 'impossible' and 1 representing 'certain'. Thus, the likelihood of most events will be assigned a numerical value in between 0 and 1; that is, a fraction or decimal. This is difficult for children, because many of them are very uncertain about fractions and decimals, and also about the relative positions of particular rational numbers on the number line. In any case, fractions being rational numbers implies that the study of probability involves ratio and proportion. There is so much research evidence which documents the difficulties pupils have with the concepts of ratio and proportion that it is impossible to do justice to it here, but Hart (1981) provides a good start to follow-up reading. It would be possible, of course, to place the likelihood of events on a six-point numerical scale, following on from the final six literal categories above, and then perhaps a ten-point scale (like marks out of ten), and this would certainly be easier than a scale from 0 to 1. However, there is no evidence which indicates how helpful this might be in the longer term.

Garfield and Ahlgren (1988) make it clear that there is considerable research evidence that a major difficulty is that ideas of probability 'appear to conflict with students' experiences and how they view the world'. Thus, the likelihood of throwing a six with a dice (this is strictly the plural, but the correct singular, 'die', is not popular with children, and is not much used by them in their own games) depends partly on how lucky the thrower is and partly on the fact that 'sixes occur less frequently than other numbers'. Similarly for adults the likelihood of meeting a vehicle coming the other way on a narrow bridge is always high, because that is how the world is. One suspects that a great many large families with children who are all of the same sex are the outcome of an expectation that, having produced three girls in succession, the probability that the next would be a boy is quite high. In fact, biologically it may be possible that the opposite is true. In coin-tossing, some children adopt particular throwing techniques because they believe that, by such means, they can at least partly control the outcomes. The 'law of averages' is much quoted when adults are hoping that 'things will even themselves out in the end'. In some ways, they might seem to, but great disappointment can follow from too rigid an adherence to this belief, for example in betting. One particularly interesting research question concerns the different intuitive beliefs held within different cultural groups around the world. Garfield and Ahlgren (1988, p. 54) quote Hawkins and Kapadia (1984), reporting on the results of dice-throwing experiments with eleven-year-old children: 'The vast majority of pupils gave mathematically sound predictions . . . [but] . . . when they found that these predictions were not confirmed by the data, many pupils reverted to past experience, hunches or cynicism.'

Once the idea of probability has been established, the next stage is most simply approached through listing and counting, and not through rules or formulae. The only way to work out the probability that the next car to come round the corner will be white is to count cars over a long enough period of time, recording white and not white. The proportion of white cars then gives us the probability. In an example like this, it is important to discuss how many data should be collected in order to make a justifiable decision, and also at what times during the week the data should be collected. The probability of throwing a 5 with an unbiased dice is 1/6, because 5 is one possible outcome of the six equally likely outcomes. The probability of throwing two 5s with two unbiased dice – one red (R) and one blue (B) – is 1/36, because there are now thirty-six likely outcomes. These are tabulated here with the number on the blue dice below the table, and number on the red dice up the left-hand side, adopting the convention of ordered pairs (or coordinates):

Number on the red dice							
6	(1, 6)	(2, 6)	(3, 6)	(4, 6)	(5, 6)	(6, 6)
5	(1, 5)	(2, 5)	(3, 5)	(4, 5)	**(5, 5)**	(6, 5)
4	(1, 4)	(2, 4)	(3, 4)	(4, 4)	(5, 4)	(6, 4)
3	(1, 3)	(2, 3)	(3, 3)	(4, 3)	(5, 3)	(6, 3)
2	(1, 2)	(2, 2)	(3, 2)	(4, 2)	(5, 2)	(6, 2)
1	(1, 1)	(2, 1)	(3, 1)	(4, 1)	(5, 1)	(6, 1)
	1	2	3	4	5	6

Number on the blue dice

Of course, any form of listing and recording will do, as long as it matches the knowledge and expertise of the pupils. From the table, we now see that only one out of the thirty-six pairs is two 5s **(5, 5)**. Likewise, the probability that, when two Smarties are to be selected from a bag containing two red, one orange and three yellow, one is red and the other is orange (a very common kind of probability question) can be seen by listing all the possibilities:

(Red 1, Red 2); **(Red 1, Orange)**; (Red 1, Yellow 1); (Red 1, Yellow 2); (Red 1, Yellow 3);

(Red 2, Orange); (Red 2, Yellow 1); (Red 2, Yellow 2); (Red 2, Yellow 3);

(Orange, Yellow 1); (Orange, Yellow 2); (Orange, Yellow 3);

(Yellow 1, Yellow 2); (Yellow 1, Yellow 3);

(Yellow 2, Yellow 3).

which gives the answer 2/15. Ultimately, and over a period of time, the notion of what might be called a theoretical probability becomes apparent. There are, of course, many situations where there is no theoretical probability, only one that can be deduced from empirical data, another example being when trying to ascertain what is the probability that a drawing pin will land point down as opposed to point up. As the study of probability progresses, and probability distributions (simply distributions of probabilities rather than frequencies) become more complicated, they are best studied with the help of appropriate computer software.

Misconceptions about probability are 'part of a way of thinking about events that is deeply rooted in most people' (Garfield and Ahlgren, 1988, p. 58). Teachers have to accept this fact, but then must adopt teaching methods which are likely to lead to better understanding. The following is a collection of practical suggestions for teaching probability and statistics based on research evidence:

- introduce topics and ideas through activities and simulations;
- confront naive intuition and errors in reasoning directly;
- emphasize qualitative reasoning, do not rush to computational methods;
- revise and reteach essential prior concepts (e.g. number line, fractions);
- relate ideas to the world outside the classroom;
- discuss misuses of statistics and probability;
- do not introduce formulae and other abstractions too soon;
- do not confuse statistical concepts by introducing probability too soon.

CHAPTER 12

Assessing mathematical attainment

INTRODUCTION

In recent years the assessment of children's mathematical achievement and attainment has become an increasingly prominent aspect of the mathematics curriculum. Although mathematical assessment and testing of children has always taken place in schools, it has now assumed a greater importance as governments and those responsible for education systems, at local and national levels, desire to obtain value for money. Although there are sound arguments for the assessment of mathematics being an integral part of what happens in schools and classrooms, teachers and mathematics educators express concern that assessment and testing, in particular, may have an undue influence on the way mathematics is taught and learned. In the remainder of this chapter assessing mathematics will be taken to include testing.

PRESSURE FOR CHANGE

For many years, governments in some countries have controlled the mathematics that was taught in their schools and the ways in which it was assessed. Other governments have left the responsibility for such decisions in the hands of teachers and mathematics educators. Over the last few years it has become apparent that the tendency internationally, is for those who determine education policy and provide the finance to implement policy, to have a greater say and, in some countries, total control over the mathematics curriculum and its associated assessment. Schools and teachers are now being asked to show that they are delivering that which has been set out by policy-makers. The demand for accountability for the mathematics which children are taught, the way it is taught and the standard of attainment which is reached is likely to be with schools for many years to come. It is incumbent upon teachers and mathematics educators to acknowledge the forces which are bringing about change in the classroom and ensure that the innovative and positive developments that have taken place over the last thirty years in the teaching and learning of mathematics are not lost in the drive to have schools, teachers, children and the mathematics curriculum measured. Assessment should be the servant of the mathematics curriculum and not its master; it should not narrow the content and the methods used to teach the subject.

THE PURPOSE OF ASSESSING CHILDREN'S MATHEMATICAL PERFORMANCE

There are many different interested parties, with their own motives and desires, who wish to see children's performance in mathematics assessed. All would agree that the major purpose for carrying out assessment of mathematical performance is formative, namely to collect data which are useful in making judgements about how children's performance can be improved. Those who receive the results of the assessing of children respond to and use the information in different ways, and may wish to influence the nature and form of future mathematics assessments in order to satisfy their own aims and needs. It is essential that the many pressures which emanate from different interested groups do not override that which teachers believe is the best for their children's mathematical development. Teachers will undoubtedly be affected by the kinds of questions which are asked of their children in mathematics tasks and tests. However, little research is available that shows how and in what ways teachers react to statutory testing of mathematics. It is too readily assumed that the content of the mathematics curriculum that is taught will be made to match what appears in tests. Although this would seem to be a reasonable assumption, research is necessary to establish what changes teachers make in both the content of their lessons and their approach to teaching mathematics as a consequence of national testing.

National assessment of children serves the purpose of informing policy-makers, at national and local levels, of whether their policies are achieving the aims which they set out. Evidence from national assessment of mathematics contributes to decisions about possible changes which may be necessary to bring about policies which will achieve the aims. There is no doubt that, in the circumstances which now prevail in many countries, the control of the curriculum is in the hands of national and local policy-makers who feel that the only worthwhile evidence on which they can act is that provided by the results of mathematics tests. They view this information as an important measure of the effectiveness of the education system. At a 'lower' level, school managers and governors, being concerned for the well-being of the children in their school, will find the outcomes of national assessment useful in informing decisions they may make for the future organization, management and resourcing of the school. Thus, a relatively poor school performance in national mathematics tests may result in an increase in funding for extra apparatus for the teaching of mathematics, or, perhaps, in a review about the adequacy of the mathematics scheme used in the school.

National testing of mathematics should be viewed positively by teachers. It is impossible for teachers with over thirty children in a class to know everything about every child's mathematical knowledge, skills and understanding. Mathematics tests may not tell teachers a great deal more than they think they know about any child, or about their class, but the possibility exists, and it would be unprofessional deliberately to ignore the outcomes of national mathematics testing.

For example, the national tests for eleven-year-olds in England and Wales have shown that many children are unable to express in writing their thinking and reasoning. In general, eleven-year-olds cannot write clearly and concisely about their mathematical thinking, giving explanations and reasons for the actions which they have performed. Thus teachers, analysing the outcomes of the national mathematics tests of their classes, will have immediately seen whether this applies to their own class and, if appropriate, have considered what implications it has for their future teaching of mathematics.

The national tests serve an important formative purpose. Planning of mathematics teaching should be based upon what children already know, what difficulties they have encountered and what they need to learn. Although it is seldom the focus of national tests in mathematics, diagnostic evidence may be gathered from the tests indicating possible error patterns which children in a class make, or misconceptions that individual children may have which had previously gone undetected. National assessment and tests are also frequently summative, in that they are used to give a child a grade or level. They may also satisfy purposes for which they were not specifically designed. Many teachers validly claim that they already know the grade or level of attainment of their children and that, because of this, the tests are redundant. This would be true if the tests did not offer other information to teachers about their children's mathematics, which they clearly do. Indeed, we would go so far as to say that, if national tests occur only with particular age groups, say seven-, eleven- and fourteen-year-olds, teachers should use all or parts of previous years' tests with the non-national tested age groups in their schools for diagnostic and planning purposes, as well as a means of measuring progress of children through the school. Mathematics assessment, including national tests, can thus serve both summative and formative functions, as well as on occasions providing early insight into children's difficulties, errors and misconceptions. At the end of the day, a mathematics test is only as good as the use to which teachers put children's responses and the aggregated outcomes.

TEACHING, LEARNING AND ASSESSING

So far, the terms 'achievement' and 'attainment' have been used together without making a distinction between them. It is helpful to define them. Children given a single task to do, say 357 + 495, will achieve success by getting the correct answer. Thus achievement will be considered to be the performance on a single task or activity. If children over a period of time are given many pairs of three-digit numbers to add, then the achievements on each of the additions may be combined in some way to decide whether they have attained the appropriate grade or level in the criterion 'addition of three-digit numbers'. Achievement is more readily measurable than attainment, as the latter involves an aggregation of the successes in the former. How, and in what way, the successes are aggregated is determined prior to a decision being made on whether a child has attained the appropriate grade or level. Thus, if twenty questions are asked targeting a criterion, then seventeen or more questions correctly answered could be the threshold for the award of the grade or level. (The minimum number of marks is referred to as the mastery level.)

Teacher assessment of attainment in aspects of mathematics frequently relies on subjective judgements based upon previous experience with children; whereas national assessment of attainment is more likely to use a predetermined rule. For example, in the 1994 National Pilot (Mathematics) for eleven-year-olds in England and Wales, the total marks in the Level 3 to 5 tests were 100. A table was included in the Teachers' Guide (SCAA, 1994) to enable teachers to work out a child's level (see Table 12.1).

Table 12.1   Level corresponding to marks achieved in the 1994 National Pilot (Mathematics) Tests

Marks	Level	Action
0–11	No level awarded	Enter for Level 1–2 Tasks
12–23	Level 2	Record Level 2
24–49	Level 3	Record Level 3
50–72	Level 4	Record Level 4
73–100	Level 5	Record Level 5
84+		Enter for Test C (Level 6)

Source: SCAA (1994)

Because teacher and national assessments of children's attainment in mathematics are based upon different approaches it is to be expected that there will be children who will not attain the same grade in the two forms of assessment. In the ENCA (Evaluation of National Curriculum Assessment) study (Shorrocks *et al.*, 1991), which looked at children's attainment of a representative sample of 395 seven-year-olds, the mathematics level outcomes of the national assessments and teacher assessments were compared in each of the areas of Number, Algebra, Shape and Space, Handling Data, and Using and Applying Mathematics. Both types of assessment were also compared with the results of tests conducted by trained assessors on individual children in a one-to-one situation. Table 12.2 shows the percentage distribution of scores across four levels in the Number Attainment Target. The ENCA report reaches the conclusion that the Kappas (a statistical measure of agreement) indicate only a slight agreement between ENCA and TA (teacher assessment), but a moderate agreement between ENCA and SAT (standard assessment test). The agreement between TA and SAT lies somewhere between. The research evidence produced by the ENCA project indicates how unlikely it is that different forms of assessment of mathematics will produce the same outcomes. This does not suggest, however, that each of the assessments has little to offer. Both teacher assessment and national testing, in whatever form, are valuable contributors to an overall picture of a child's attainment in mathematics.

In most countries children, in their last year of schooling, take a terminal

Table 12.2   Percentage distribution of scores across Level W, 1, 2 and 3 of Teacher Assessment (TA), National Tests (SAT) and ENCA tests

	W	1	2	3	Kappa
			Levels		
TA	5	31	61	3	TA: SAT = 0.35
SAT	3	54	32	11	TA: ENCA = 0.20
ENCA	2	67	28	2	SAT: ENCA = 0.42

mathematics examination. Very often this is the only national assessment in mathematics some children experience in their whole life in school. In England and Wales, children are currently assessed on a national scale four times during their compulsory years in school, at the ages of seven, eleven, fourteen and sixteen years. Such a concentration of assessment and testing is highly likely to exert an influence on teachers and children. Testing procedures suggest to teachers and children what is important in the teaching and the learning of mathematics. If a national test contains questions, for example, devoted solely to number, teachers will, quite rightly, place greater emphasis in their teaching on number than on, say, geometry. Children experiencing such tests would not take kindly to teachers spending time on investigations with polygons if they did not appear in the tests. There is a responsibility on those who develop national tests, and upon those who advise and guide them, to attempt to ensure that the tests fully reflect not only the wide mathematical content that is in a national curriculum, but also that the nature and mode of assessment matches the philosophy of learning and teaching mathematics which occurs in classrooms.

However, teachers' beliefs about the nature of mathematics, and about how children learn and should be taught the subject, vary widely, even in the same school. Consequently, a variety of teaching styles is to be found in operation in mathematics lessons, each reflecting the belief of the teacher about mathematics as a subject to be learned in school. Agencies that develop tests are unable to match every possible teaching style and, therefore, cannot satisfy the wishes of every teacher. Teachers who view mathematics as a collection of knowledge and skills will be disappointed in a test which uses mainly open questions, concentrating on assessing process aspects of mathematics (see Chapter 4). At the other extreme, teachers who favour an investigative approach to teaching and learning would not wish to see a mathematics test dominated by questions which tested only algorithmic skills in number. Which of the questions in Figure 12.1 would you wish to see used in a national test to assess addition of three single-digit numbers? What are the reasons for your choice? What is it about the other questions which made you not choose them? Teachers should constantly analyse national tests to see in what ways, if at all, what they teach and the approaches they use match the tests. This is not to suggest that teachers should expect to find a match, but that any differences which are apparent indicate that a review of what happens in the classroom should be undertaken and/or those responsible for the tests should be made aware of the mismatch. Remember, it may be the tests which need modifying to fall in line with what is taught in schools, not the reverse. We should seek to be pro-active, searching for modes of assessment that reflect the best in the teaching

Question 1 $3 + \square + 5 = 10$

Question 2 Find the missing number which makes the
sum of the numbers down and across equal 10

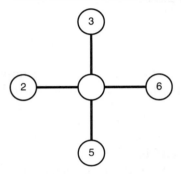

Question 3 Find three numbers which add up to 10

$\square + \square + \square = 10$

Figure 12.1   Assessing simple addition

of mathematics, not changing what takes place in classrooms to match the worst of what can be in written tests.

Mathematics assessment, whether by the teacher or by external tests, can serve many purposes for schools and teachers. Schools wishing to keep parents informed about their child's progress need reliable evidence of children's performance. Teacher assessment is more likely to provide, for parents, a detailed breakdown of a child's attainment in different aspects of mathematical content: number, algebra, shape and space, and handling data. The outcomes of national tests are more likely to provide an overall picture of a child's attainment in mathematics, not broken down into its many facets. Both national tests and teacher assessment can assist teachers in monitoring the mathematical progress of each child and furnish information about the performance of a class through its advancement from year to year. The mathematics coordinator, or head of department, should keep records of class performance for each year group and, towards the end of each year, review the results of the end-of-year assessment in the light of the results of previous years. A more detailed record of the mathematical attainment of each individual child should be kept as they progress through the school. Such records reveal when children cease to make the anticipated development, particularly in relation to their peers. Annual reviews of the attainment of individual children should be an essential feature of the evaluation of a mathematics curriculum. Evidence for such reviews may be obtained from teacher assessment, national testing and testing using teacher-made or commercial mathematics tests. Information from assessment of children can also contribute to teachers' evaluation of their own performance in implementing the mathematics curriculum. The strengths and weaknesses of the children are an indicator of which aspects of the curriculum have been taught, but not necessarily learned. For example, if a large number of children in a class produce the following subtraction:

$$
\begin{array}{r}
4\,3 \\
-1\,7 \\
\hline
3\,4
\end{array}
$$

then a teacher should begin to question the method of teaching that was used.

Larcombe (1985) categorizes the aims of assessment, from the perspective of teachers, and describes the purpose of each, together with when and how assessment might be carried out (see Table 12.3). This is potentially an overwhelming list of what teachers should be doing in their schools and classrooms, yet all four types of assessment have important roles in the teaching and learning of mathematics. Each school must make its own decisions about its assessment priorities

Table 12.3   The aims of assessment

Type of assessment	What purposes might assessment serve?	When might assessment be done?	How might assessment be carried out?
Type 1: to compare pupil's ability or performance with other pupils'	Producing rank orders. Aid to setting/ grouping. Screening to identify pupils needing closer observation, and diagnosis at both ends of the ability range	At any time across a group of pupils	Formal examination or test situation for whole group. Allocation of marks on scale for routine marking of classwork and homework
Type 2: to identify areas of mis-understanding, weakness or failure	Diagnosis of weakness. Inform-ation for planning remedial action	After previous screening or referral has indicated need for diagnostic testing	Formal testing individually. Informal testing individually. Analysis of routine marking of classwork and homework
Type 3: to assess whether maths previously encountered has been understood	Monitoring progress and achievement. Information for planning future work. Evaluation of use-fulness of work given	After previous screening or referral has indicated need for diagnostic testing. Regularly on com-pletion of sections of work.	Formal testing in-dividually. Informal testing individually. Observation of pupil
Type 4: to reveal other facets of the pupil's develop-ment which might have some bearing upon future learning	To provide background information for course planning. Evaluation of impact of work upon pupil	Continuously as pupil works at mathematics	Observation of pupil. Informal testing. Completion of check-lists

Source: Larcombe (1985)

and develop the organization which provides the opportunity for successful implementation of a school policy in the assessment and testing of children's mathematical performance.

THE 'WHAT' AND 'HOW' OF ASSESSING MATHEMATICS

There is a clear need for a detailed and categorical definition of what in a curriculum is to be assessed. Is it only the content which should be assessed, or should creativity, insight and investigative process skills also be assessed?

There have been many attempts to describe the nature of mathematics in terms of assessable objectives (Avital and Shettleworth, 1968; Wood, 1968; Hollands, 1972). Some teachers and mathematics educators would claim that it is not possible to produce such categorizations without distorting the true nature of the subject. There are those who would vehemently assert that it is not possible to assess children's ability to investigate mathematical process problems and strategies, as it is contrary to the spirit of investigations and investigative work. Much work has yet to be done on the development of assessment materials which reflect every aspect of content and approach to the learning of mathematics. Of those taxonomies referred to above, Hollands' comes the closest to the desires of those who believe there is more to mathematics assessment than content. He lists nine categories which teachers may find useful when considering what mathematics should be assessed.

1 Basic terms and facts
2 Techniques and manipulative skills
3 Understanding, first level
4 Application
5 Analysis
6 Synthesis
7 Creativity
8 Understanding, second level
9 Evaluation

For a more detailed description of each category, with illustrative examples, reference to Hollands' article is advised. Teachers could use the categories as a starting point for assessing children, adding to the list other aspects of mathematics which they feel are omitted and should be assessed as they contribute to a fuller picture of a child's attainment. If assessment develops out of the curriculum, then the content of the assessment, whether teacher assessment or national testing, should reflect the purposes of assessment, a view of the nature of mathematics and what is known about how children learn the subject. It is much easier to see if national tests satisfy these three principles, as they are open to teacher scrutiny. The same cannot be claimed for teacher assessment, as this tends to take place in the privacy of a classroom. Teachers must not allow themselves to fall into the trap of allowing their prejudices about mathematics to influence their assessment of children. They, too, should ensure that their assessments truly reflect the curriculum in all its different facets of content and approach and do not over-emphasize one aspect of mathematics, say number, to the detriment of others, such as shape and space, or algebra.

Accurate teacher assessment can be achieved only if it is informed by knowledge and awareness of children's mathematical performance over a period of time involving both informal and formal assessment procedures. In recent years greater emphasis has been placed upon informal assessment of children in a normal classroom environment. Teacher observation of children working on mathematical activities, and discussing with other children the methods they could use and the outcomes achieved, has been a substantial contributory factor in the overall assessment of children in primary schools. This approach, whilst considered to be manageable in the context of classroom organization, has major disadvantages. It not only relies on teachers having the expertise to recognize appropriate behaviour worthy of accreditation towards a grade or level, but requires that they identify it when it occurs in what often is a learning experience for the children.

Observation of children is usually supported by teachers building up, for each child, a portfolio of work which is representative of their normal performance. A portfolio is a collection of children's work which may include pages of calculations, assignments, projects and investigations, as well as reports of mathematical activities which have been completed as part of a child's normal classroom experiences. Reliable assessment of children requires evidence that children can perform consistently at different times and in a variety of contexts. If portfolios are used to assist teachers in their assessment, then it is essential that criteria be established which enable teachers to select work which accurately represents children's normal performance. It is all too easy to be excited by a very good piece of work produced by a child and include it in a portfolio when the work, for that child, is extraordinary and not truly representative.

As the mathematics curriculum covers many diverse topics, such as addition of natural numbers and the construction of shapes, portfolios should contain children's work which covers both as many areas of mathematics as is manageable and also the categories of objectives listed by Hollands (1972). A piece of work which illustrates a child's achievement in the addition of pairs of two-digit numbers says nothing about the child's achievement in subtraction of two-digit numbers. Yet it is impossible to include in a portfolio evidence from every small area of the curriculum. Thus, a selection of areas should be made for each age group which a school believes are representative of similar areas, but which for the sake of manageability are omitted. Such selections should not be left to individual teachers, but be part of a school mathematics assessment policy. In this way, assessment from year to year achieves some form of consistency.

Further evidence for teacher assessment can be obtained from classroom tests which are given at regular intervals throughout a year. The outcomes of tests can contribute towards the overall teacher assessment of children's mathematical attainment, and the tests can be chosen or devised by teachers to fit into the current practice of the class. If classroom tests match the mathematical activities with which children have been working, then they are viewed by children as an integral part of their learning of mathematics. If the outcomes of the tests are also used to diagnose children's misconceptions and errors, then children perceive the tests as helping them overcome problems in the learning of mathematics. Written tests can be devoted to a particular and narrow area of mathematics; for example, the drawing of graphs or the recognition of polygons. They can also be more extensive in their assessment, covering more than one topic in mathematics. The latter type of test provides teachers with evidence of aspects of the subject which may not have been retained, indicating the need for revision. The value of such tests should not be underrated as, when new areas of mathematics are taught, they build upon previous knowledge and understanding. There is always the danger that tests test only that which is easy to test, usually knowledge and skills. This does not have to be the case and, if used with thought and planning, tests can be of immeasurable benefit to teachers in their quest to produce accurate and reliable teacher assessment of their children.

 Teacher-made tests are very personal and can be made to match individual assessment needs. The results from such tests are immediately available to the teacher and children, and the interpretation of children's responses can be done more accurately by a teacher. However, the compiling of tests is far from easy, and there is much to consider when writing tests, even ones apparently as simple as a test of ten questions on 'addition of pairs of single-digit numbers'. Every test should have a clear mathematical assessment objective, like the one just described. But how is the test to be presented in order to assess children validly? How are children expected or allowed to respond, and in what form? Will they be allowed to use apparatus, say cubes or counters, or can they use fingers? Which pairs of single-digit numbers should be used to make up the ten addition questions? Here is a test of ten such questions we have written without much thought given to the test's content or structure.

1) $5 + 8 =$	2) $3 + 9 =$
3) $4 + 2 =$	4) $8 + 4 =$
5) $6 + 2 =$	6) $2 + 7 =$
7) $7 + 5 =$	8) $6 + 5 =$
9) $9 + 2 =$	10) $6 + 4 =$

Look at the test we have written. Analyse the frequency of the numbers we have used. Have we omitted any numbers? Do some numbers occur more frequently than others? What about the answers? Do the answers have a reasonable coverage of all possible answers? This very 'simple' test illustrates some of the problems that confront those who wish to write their own tests. Also the test just described is 'closed', in that each question has a unique answer. Some teachers would wish to assess the same mathematics using a more 'open' approach as in the following set of ten questions:

Fill in the missing numbers

1) ☐ + ☐ = 8   2) ☐ + ☐ = 14

3) ☐ + ☐ = 15   4) ☐ + ☐ = 13

5) ☐ + ☐ = 12   6) ☐ + ☐ = 6

7) ☐ + ☐ = 7   8) ☐ + ☐ = 9

9) ☐ + ☐ = 19   10) ☐ + ☐ = 17

A small minority of teachers would complain that even this test is not suffi-ciently open and would prefer ten questions identical to this question.

Write an addition.

☐ + ☐ = ☐

All three tests assess what Hollands (1972) categorized as 'Basic skills', but the latter two tests also assess some of the other objectives. Which of the three forms of test do you feel matches the way your children are accustomed to responding? Would the other two forms be a fair test of your children's achievements in single-digit addition?

A teacher-made test of this aspect of mathematics is usually devised to match how children have learned. Others would argue, however, that if children know and understand addition of single-digit numbers, then they should be able to respond successfully whatever the form of the questions. This is a view which an agency developing national tests may take, as, however the questions are framed, they will inevitably not match the way some children have been taught.

Teachers should seriously consider the use of commercial tests to provide them with evidence for teacher assessment, if such tests satisfy the assessment objec-tives which the teacher has targeted and the results are set in the context of information obtained in different ways. Commercial tests are usually standard-ized, in that they have been administered under strict uniform conditions to a large, representative sample of children. A standardized test may be norm-referenced, in that the referencing is made to the distribution of the scores of the large sample. A norm-referenced test could be devised on a narrow area of the curriculum, such as the naming of polygons, or 'long multiplication'. In such cases it is advisable to have a large number of questions which are spread 'equally' over the domain of difficulty, from the very easy to the very difficult. This ensures that there will be a variation in the scores of the sample of children. Thus children taking the test at a later time may have their scores judged in comparison to the normed sample. Norm-referenced tests enable teachers to discriminate between children, and a child's performance is judged relative to other children.

Standardized tests may, however, be criterion-referenced in that questions target a specific mathematical criterion or criteria, related to a particular area or areas of mathematics. A criterion could be 'use place value with numbers up to 1000'. One approach to a criterion-referenced test, targeting a single criterion, is to develop a set of questions which spreads over the domain of difficulty of the criterion. A mastery level is then determined which is an indicator of a child's attainment of the criterion. The level of mastery is an arbitrary decision, but is usually in the range 75 to 85 per cent of the possible marks. The advantage of criterion-referenced tests lies in their ability to inform teachers what their children can and cannot do.

DIAGNOSTIC ASSESSMENT

Diagnostic assessment is a type of formative assessment, as its function is to provide teachers with information about children's difficulties, misconceptions and errors. Information of this kind is essential if teachers are to design appropriate activities for their children, building upon a stable base of understanding. Norm- and criterion-referenced tests can provide teachers with limited diagnostic information, although that is not their major focus. Diagnostic data can be obtained from the way children respond to questioning in the classroom, and to their work products. This is likely to be only the beginning of a diagnostic process, which should follow the identification of an error or misconception. For example, if a child is asked, 'How many sides has a triangle?' and responds with the answer 'four', then a lack of knowledge about triangles has been identified. A twelve-year-old boy showed the authors what a half was by drawing the square shown in Figure 12.2. The reason why he partitioned the square into four parts was not immediately obvious. It appeared that he was able to represent one half successfully, using the region aspect of fractions (see Chapter 8). When asked to show a third he responded with the same partitioning, but shaded three parts (see Figure 12.3). The movement of the boy's hands helped us to understand the interpretation which he was giving to each of the parts of the written representation of a third ($^1/_3$). His hand swept around the whole square (the '1'), and then pointed to the three shaded parts (the '3'), as he said 'You see – this is one third'. We were unable to find out why the boy divided the square into four 'equal' parts on both occasions, as he left the room immediately the break bell sounded. Just prior to the interview the child had been multiplying proper fractions successfully but, understandably, had failed to multiply mixed numbers correctly. What brought about this state of affairs should be clearly established before activities are devised to remedy the situation. Too often teachers, before establishing the root cause of an error or misconception, reteach a child using the same or similar activities, in the belief that more of the same will bring about the requisite knowledge and understanding. Remedial activities should be developed in the light of evidence of the causes of the original error or misconception.

A diagnostic approach to teaching mathematics relies upon diagnostic assessment of children's misconceptions or errors. At present there are no intensive diagnostic assessment instruments available for teachers to use when they are

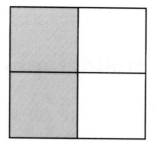

Figure 12.2   One-half: a child's correct response

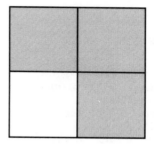

Figure 12.3   One-third: a child's incorrect response

confronted by children having difficulties. Teachers, therefore, can rely only upon observation, questioning and work products to help them with the identification and subsequent causes of misconceptions and errors. There appears to be little diagnostic expertise amongst mathematics teachers in schools to provide the support individual teachers need when they have children with mathematical difficulties which they are unable to identify. This is an aspect of the assessment of mathematics in schools where considerable work remains to be done.

Bibliography

Assessment of Performance Unit (1980a) *Mathematical Development: Primary Survey Report Number 1*. London: HMSO.

Assessment of Performance Unit (1980b) *Mathematical Development: Secondary Survey Report Number 1*. London: HMSO.

Assessment of Performance Unit (undated) *Mathematical Development: A Review of Monitoring in Mathematics 1978 to 1982*. Slough: NFER.

Association of Teachers in Colleges and Departments of Education (1967) *Teaching Mathematics: Main Courses in Mathematics in Colleges of Education*. Cambridge: Cambridge University Press.

Association of Teachers of Mathematics (1969) *Notes on Mathematics in Primary Schools*. Cambridge: Cambridge University Press.

Avital, S. M. and Shettleworth, S. J. (1968) *Objectives for Mathematics Learning*. Toronto: Ontario Institute for Studies in Education.

Ball, G. (1990) *Talking and Learning: Primary Maths for the National Curriculum*. Oxford: Basil Blackwell.

Banwell, C. S., Saunders, K. D. and Tahta, D. G. (1972) *Starting Points*. Oxford: Oxford University Press.

Barnard, T. and Saunders, P. (1994) Superior sums that don't add up to much. *Guardian*, 28 December.

Baroody, A. J. (1987) *Children's Mathematical Thinking*. New York: Teachers College Press.

Bideaud, J., Meljac, C. and Fischer, J. (eds) (1992) *Pathways to Number: Children's Developing Numerical Abilities*. Hillsdale, NJ: Erlbaum.

Booth, L. R. (1984) *Algebra: Children's Strategies and Errors*. Windsor: NFER-Nelson.

Boulton-Lewis, G. M. and Halford, G. S. (1990) An analysis of the value and limitations of mathematical representation used by teachers and young children. In G. Booker, P. Cobb and T. de Mendicuti (eds), *Proceedings of the Fourteenth International Conference for the Psychology of Mathematics Education*. Mexico: CONACYT.

Brown, M. (1981) Number operations. In K. M. Hart (ed.), *Children's Understanding of Mathematics: 11–16*. London: John Murray.

Brown, S. I. (1984) The logic of problem generation: from morality and solving to deposing and rebellion. *For the Learning of Mathematics*, 4(1), 9–20.

Brown, S. I. and Walter, M. I. (1983) *The Art of Problem Posing*. Philadelphia: Franklin Institute Press.

Brownell, W. (1942) Psychological considerations in the learning and teaching of arithmetic. In W. D. Reeve (ed.), *The Teaching of Arithmetic*. New York: Teachers College Press.

Bruner, J. S. (1960) *The Process of Education*. Cambridge, MA: Harvard University Press.

Burton, L. (1994) *Children Learning Mathematics: Patterns and Relationships*. Hemel Hempstead: Simon & Schuster.

Carr, K. and Katterns, B. (1984) Does the number line help? *Mathematics in School*, **13**(4), 30–4.

Carraher, T. N. (1985) The decimal system: understanding and notation. In L. Streefland (ed.), *Proceedings of the Ninth International Conference for the Psychology of Mathematics Education*. Utrecht: University of Utrecht.

Cobb, P. and Wheatley, G. (1988) Children's initial understandings of ten. *Focus on Learning Problems in Mathematics*, **10**(3), 1–28.

Cockcroft, W. H. (1982) *Mathematics Counts*. London: HMSO.

Collis, K. F. (1975) *A Study of Concrete and Formal Operations in School Mathematics*. Hawthorn, Victoria: Australian Council for Educational Research.

Cresswell, M. and Gubb, J. (1987) *The Second International Mathematics Study in England and Wales*. Windsor: NFER-Nelson.

Curcio, F. R. (1987) Comprehension of mathematical relationships expressed in graphs. *Journal for Research in Mathematics Education*, **18**(5), 382–93.

Department of Education and Science (1979) *Mathematics 5–11: A Handbook of Suggestions*. London: HMSO.

Department of Education and Science (1985) *Mathematics from 5 to 16*. London: HMSO.

Department of Education and Science (1988) *Mathematics for Ages 5–16*. London: HMSO.

Department of Education and Science (1991) *Mathematics in the National Curriculum*. London: HMSO.

Dickson, L., Brown, M. and Gibson, O. (1984) *Children Learning Mathematics*. Eastbourne: Holt, Rinehart & Winston.

Dienes, Z. P. (1960) *Building Up Mathematics*: London: Hutchinson Educational.

Driscoll, M. (1982) *Research within Reach: Secondary School Mathematics*. Reston, VA: National Council of Teachers of Mathematics.

Ernest, P. (1991) *The Philosophy of Mathematics Education*. Basingstoke: Falmer Press.

Fisher, R. and Vince, A. (1989) *Investigating Maths*. Oxford: Basil Blackwell.

Fitzgerald, A. F. (1981) *Mathematics in Employment (16–18)*. Bath: University of Bath.

Frobisher, L. J. (1994) Problems, investigations and an investigative approach. In A. Orton and G. Wain (eds), *Issues in Teaching Mathematics*. London: Cassell.

Frobisher, L. J., MacNamara, A. and Threlfall, J. (1993) Children's responses to addition facts using a computer. Unpublished paper, University of Leeds.

Fuson, K. C. (1982) An analysis of the counting-on solution procedure in addition. In T. P. Carpenter, J. M. Moser and T. A. Romberg, *Addition and Subtraction: A Cognitive Perspective*. Hillsdale, NJ: Erlbaum.

Fuson, K. C. (1988) *Children's Counting and Concepts of Number*. New York: Springer-Verlag.

Fuson, K. C. and Hall, J. W. (1983) The acquisition of early number word meanings: a conceptual analysis and review. In H. P. Ginsburg (ed.), *The Development of Mathematical Thinking*. New York: Academic Press.

Garfield, J. and Ahlgren, A. (1988) Difficulties in learning basic concepts in probability and statistics: implications for research. *Journal for Research in Mathematics Education*, **19**(1), 44–63.

Gelman, R. and Gallistel, C. R. (1978) *The Child's Understanding of Number*. Cambridge, MA: Harvard University Press.

Gibbs, W. and Orton, J. (1994) Language and mathematics. In A. Orton and G. Wain (eds), *Issues in Teaching Mathematics*. London: Cassell.

Ginsburg, H. (1977) *Children's Arithmetic: The Learning Process*. New York: Van Nostrand.

Grouws, D. A. and Good, T. L. (1989) Issues in problem-solving instruction. *Arithmetic Teacher*, **36**(8), 34–5.

Harling, P. (1990) *100s of Ideas for Primary Maths: A Cross-curricular Approach*. London: Hodder & Stoughton.

Hart, K. (1981) Ratio and proportion. In K. M. Hart (ed.), *Children's Understanding of Mathematics: 11–16*. London: John Murray.

Hart, K. (1989) There is little connection. In P. Ernest (ed.), *Mathematics Teaching: The State of the Art*. Lewes: Falmer Press.

Hawkins, A. S. and Kapadia, R. (1984) Children's conceptions of probability – A psychological and pedagogical review. *Educational Studies in Mathematics*, **15**, 349–78.

Herscovics, N. (1989) Cognitive obstacles encountered in the learning of algebra. In S. Wagner and C. Kieran (eds), *Research Issues in the Learning and Teaching of Algebra*. Reston, VA: National Council of Teachers of Mathematics/Lawrence Erlbaum Associates.

Herscovics, N. and Kieran, C. (1980) Constructing meaning for the concept of equation. *Mathematics Teacher*, **73**, 572–81.

Hiebert, J. (1988) A theory of developing competence with written mathematical symbols. *Educational Studies in Mathematics*, **19**(3), 333–55.

Hollands, R. D. (1972) Educational technology: aims and objectives in teaching mathematics. *Mathematics in School*, **1**(2), 23–4; **1**(3), 32–3; **1**(5), 20–1; **1**(6), 22–3.

Holmes, P. (1980) *Teaching Statistics 11–16*. Slough: Foulsham Educational.

Hughes, M. (1981) Can pre-school children add and subtract? *Educational Psychology*, **1**(3), 207–49.

Hughes, M. (1986) *Children and Number*. Oxford: Basil Blackwell.

Janvier, C. (1989) Representation and contextualization. In G. Vergnaud, J. Rogalski and M. Artigue (eds), *Proceedings of the Thirteenth International Conference for the Psychology of Mathematics Education*. Paris: CNRS.

Johnson, D. A. and Rising, G. R. (1967) *Guidelines for Teaching Mathematics*. Belmont, CA: Wadsworth.

Kerslake, D. (1979) Visual mathematics. *Mathematics in School*, **8**(2), 34–5.

Kerslake, D. (1981) Graphs. In K. M. Hart (ed.), *Children's Understanding of Mathematics: 11–16*. London: John Murray.

Kilpatrick, J. (1987) Problem formulating: where do good problems come from? In A. H. Schoenfeld (ed.) *Cognitive Science and Mathematics Education*. Hillsdale, NJ: Erlbaum.

Kirkby, D. (1989) *Spectrum Maths*. London: Unwin Hyman.

Kirkby, D. and Patilla, P. (1987) *Maths Investigations*. London: Hutchinson.

Kouba, V. L., Carpenter, T. P. and Swafford, J. O. (1989) Number operations. In M. M. Lindquist (ed.), *Results from the Fourth Mathematics Assessment of the National Assessment of Educational Progress*. Reston, VA: National Council of Teachers of Mathematics.

Küchemann, D. (1981a) Algebra. In K. M. Hart (ed.), *Children's Understanding of Mathematics: 11–16*. London: John Murray.

Küchemann, D. (1981b) Reflections and rotations. In K. M. Hart (ed.), *Children's Understanding of Mathematics: 11–16*. London: John Murray.

Laborde, C. (1990) Language and mathematics. In P. Nesher and J. Kilpatrick (eds), *Mathematics and Cognition*. Cambridge: Cambridge University Press.

Larcombe, A. (1985) *Mathematical Learning Difficulties in the Secondary School*. Milton Keynes: Open University Press.

Lerman, S. (1989) Investigations: where to now? In P. Ernest (ed.), *Mathematics Teaching: The State of the Art*. Lewes: The Falmer Press.

Lesh, R., Landau, M. and Hamilton, E. (1980) Rational number ideas and the role of representational systems. In R. Karplus (ed.), *Proceedings of the Fourth International Conference for the Psychology of Mathematics Education*. Berkeley: University of California.

Lindquist, M. M. and Shulte, A. P. (eds) (1987) *Learning and Teaching Geometry, K-12*. Reston, VA: National Council of Teachers of Mathematics.

MacKernan, J. (1982) The merits of verbalism. *Mathematics in School*, **11**(4), 27–30.

Mason, J., Graham, A., Pimm, D. and Gowar, N. (1985) *Routes to/Roots of Algebra*. Milton Keynes: Open University Press.

Matz, M. (1983) Towards a computational theory of algebraic competence. *Journal of Mathematical Behaviour*, **3**, 93–166.

Mirua, I. T. and Okamoto, Y. (1989) Comparisons of US and Japanese first graders' cognitive representation of number and understanding of place value. *Journal of Educational Psychology*, **81**(1), 109–13.

Mitchelmore, M. C. (1978) Developmental stages in children's representation of regular solid figures. *Journal of Genetic Psychology*, **133**, 229–39.

Moses, B., Bjork, E. and Goldenberg, E. P. (1990) Beyond problem solving: problem posing. In T. J. Cooney and C. R. Hirsch (eds), *Teaching and Learning Mathematics in the 1990s*. Reston, VA: National Council of Teachers of Mathematics.

Murray, J. (1939) The relative difficulty of the basic number facts. In The Scottish Council for Research in Education (eds) *Studies in Arithmetic*, Vol. 1. London: University of London Press.

National Council of Teachers of Mathematics (1980) *An Agenda for Action: Recommendations for School Mathematics of the 1980s*. Reston, VA: NCTM.

National Council of Teachers of Mathematics (1989) *Curriculum and Evaluation Standards for School Mathematics*. Reston, VA: NCTM,

National Curriculum Council (1989) *Mathematics Non-Statutory Guidance*. York: NCC.

Nuffield Mathematics Project (1967a) *I Do, and I Understand*. London: Chambers/Murray/Wiley.

Nuffield Mathematics Project (1967b) *Pictorial Representation*. London: Chambers/Murray/Wiley.

Nuffield Maths 5–11 (1979) *Bronto Books*. London: Longman.

Orton, A. (1992) *Learning Mathematics: Issues, Theory and Classroom Practice* (2nd edition). London: Cassell.

Orton, A. (1994) Learning mathematics: implications for teaching. In A. Orton and G. Wain (eds), *Issues in Teaching Mathematics*. London: Cassell.

Orton, A. and Orton, J. (1994) Students' perception and use of pattern and generalization. In J. P. da Ponte and J. F. Matos (eds) *Proceedings of the Eighteenth International Conference for the Psychology of Mathematics Education*. Lisbon: University of Lisbon.

Orton, J. E. H. (1989) A study of some aspects of children's perception of pattern in relation to shape. University of Leeds: Unpublished MEd thesis.

Orton, J. (1994) Pattern in relation to shape. In G. T. Wain (ed.), *Research Papers of the British Congress on Mathematical Education*. Leeds: University of Leeds Centre for Studies in Science and Mathematics Education.

Pimm, D. (1987) *Speaking Mathematically*. London: Routledge & Kegan Paul.

Pollard, M., with Boucher, J., Shuard, H. and Wain, G. (1983) *Today's World Series*. Eastbourne: Holt, Rinehart & Winston.

Polya, G. (1957) *How to Solve It*. New York: Doubleday Anchor.

Resnick, L. B. (1983) A developmental theory of number understanding. In H. P. Ginsburg (ed.), *The Development of Mathematical Thinking*. New York: Academic Press.

Richards, J. (1991) Mathematical discussions. In E. von Glasersfeld (ed.), *Radical Constructivism in Mathematics Education*. Dordrecht: Kluwer.

Robitaille, D. F. and Garden, R. A. (1989) *The IEA Study of Mathematics 2: Contexts and Outcomes of School Mathematics*. Oxford: Pergamon Press.

Russell, R. L. and Ginsburg, H. P. (1984) Cognitive analysis of children's mathematical difficulties. *Cognition and Instruction*, 1(2), 217–44.

Scandura, J. M. (1971) *Mathematics: Concrete Behavioral Foundations*. New York: Harper & Row.

School Curriculum and Assessment Authority (1994) *National Pilot Teacher's Guide*. London: SCAA.

School Examinations and Assessment Council (1988) *APU Mathematics Monitoring 1984–88 (Phase 2)*. London: HMSO.

Shorrocks, D., Daniels, S., Frobisher, L., Nelson, N., Waterson, A. and Bell, J. (1991) *The Evaluation of National Curriculum Assessment at Key Stage 1*. Leeds: University of Leeds.

Shuard, H. (1986) *Primary Mathematics Today and Tomorrow*. York: Longman.

Shuard, H., Walsh, A., Goodwin, J. and Worcester, V. (1991). *PrIME: Calculators, Children and Mathematics*. Hemel Hempstead: Simon & Schuster.

Skemp, R. R. (1971) *The Psychology of Learning Mathematics*. Harmondsworth: Penguin Books.

Skemp, R. R. (1976) Relational understanding and instrumental understanding. *Mathematics Teaching*, 77, 20–6.

Starkey, P. and Gelman, R. (1982) The development of addition and subtraction abilities prior to formal schooling in arithmetic. In T. P. Carpenter, J. M. Moser and T. Romberg (eds), *Addition and Subtraction: A Cognitive Perspective*. Hillsdale, NJ: Erlbaum.

Steffé, L. P., von Glasersfeld, E., Richards, J. and Cobb, P. (1983) *Children's Counting Types: Philosophy, Theory and Application*. New York: Praeger.

Straker, A. (1993) *Talking Points in Mathematics*. Cambridge: Cambridge University Press.

Sutherland, R. (1990) The changing role of algebra in school mathematics: the potential of computer-based environments. In P. Dowling and R. Noss (eds), *Mathematics versus the National Curriculum*. Basingstoke: Falmer Press.

Tall, D. and Thomas, M. (1991) Encouraging versatile thinking in algebra using the computer. *Educational Studies in Mathematics*, **22**, 125–47.

Thorburn, P. and Orton, A. (1990) One more learning difficulty. *Mathematics in School*, **19**(3), 18–19.

Threlfall, J., Frobisher, L. J. and MacNamara, A. (1995) Inferring the use of recall in simple addition. *British Journal of Educational Psychology* **65**, 425–39.

Wagner, S. (1983) What are these things called variables? *Mathematics Teacher*, **76**, 474–9.

Walter, M. (1981) Do we rob students of a chance to learn? *For the Learning of Mathematics*, **1**(3), 16–18.

Ward, M. (1979) *Mathematics and the 10-year-old*. London: Evans.

Watson, F. R. (1983) Investigation. *Educational Analysis*, **5**(3), 33–44.

Wilder, R. L. (1978) *Evolution of Mathematical Concepts*. Milton Keynes: Open University Press.

Wood, R. (1968) Objectives in the teaching of mathematics. *Educational Research*, **10**, 83–98.

Name index

Subject index